Gas Infra-red, the Green Heat

History of IR heating and Infra-red Heating Design and Economic Analysis

RIVER PUBLISHERS SERIES IN ENERGY MANAGEMENT

Series Editor:

MICHELE ALBANO
Aalborg University, Denmark

The "River Publishers Series in Energy Management" is a series of comprehensive academic and professional books focussing on management theory and applications for energy related industries and facilities. Books published in the series serve to provide discussion and exchange information on management strategies, techniques, methodologies and applications, with a focus on the energy industry.

Topics include management systems, handbooks for facility management, safety, security, industrial strategies, maintenance and financing, impacting organizational communications, processes and work practices. Content is also featured for energy resilient and high-performance buildings.

The main aim of this series is to serve as a useful reference for academics, researchers, managers, engineers, and other professionals in related matters with energy management practices.

Books published in the series include research monographs, edited volumes, handbooks and textbooks. The books provide professionals, researchers, educators, and advanced students in the field with an invaluable insight into the latest research and developments.

Topics covered in the series include, but are not limited to:

- Facility management;
- Safety and security;
- Management systems and solutions;
- Industrial energy strategies;
- Financing and costs;
- Energy resilient buildings;
- Green buildings management.

For a list of other books in this series, visit www.riverpublishers.com

Gas Infra-red, the Green Heat

History of IR heating and Infra-red Heating Design and Economic Analysis

George Kritzler

USA

River Publishers

Routledge
Taylor & Francis Group
NEW YORK AND LONDON

Published 2025 by River Publishers
River Publishers
Broagervej 10, 9260 Gistrup, Denmark
www.riverpublishers.com

Distributed exclusively by Routledge
605 Third Avenue, New York, NY 10017, USA
4 Park Square, Milton Park, Abingdon, Oxon OX14 4RN

Gas Infra-red, the Green Heat / by George Kritzler.

Routledge is an imprint of the Taylor & Francis Group, an informa business.

ISBN 978-87-7004-657-2 (hardback)
ISBN 978-87-7004-786-9 (paperback)
ISBN 978-87-7004-659-6 (online)
ISBN 978-87-7004-658-9 (master ebook)

DOI: 10.1201/9788770046589

Contents

Preface

This book is intended for anyone interested in designing energy-efficient comfort heating systems using gas-fired infra-red heaters. Infra-red heating is truly the "greenest" of heating systems available for "high bay" applications and in some cases even "low bay" areas.

Infra-red was discovered in 1800 by an English astronomer Sir William Herschel. Sir Herschel performed some basic experiments. By measuring the heat of sun light diffused through a prism, he determined that blue light provided the least heat and that the temperature gradually increased as the color changed to the red spectrum. When he continued beyond the visible red spectrum, he was astonished to learn that the highest temperature he was able to achieve was in the invisible spectrum–which is the infra-red spectrum.

Infra-red is not new. Obviously it has been around for as long as our planet has been orbiting the sun. The enormous amount of heat that the earth receives is transmitted from the sun in the form of infra-red, along with visible light. It has been estimated that the temperature of the sun is about $5{,}800\,^{\circ}\mathrm{C}$, which means that it is radiating at approximately $2{,}600{,}000\,\mathrm{BTUH/ft^2}$. However, most of this energy is given off as visible light due to the extremely high source temperature. Since the earth is so far away, it receives only a small amount of the infra-red emitted by the sun. The intensity of the radiant heat throughout the United States amounts to only $200–250\,\mathrm{BTUH/ft^2}$.

Infra-red is a small part of the electromagnet spectrum and is located between visible light at the top end of the radar and microwave portion of the electromagnetic spectrum. The source temperature, and the fact that the human eye cannot see low-temperature infra-red, is the main difference between visible light and infra-red energy. The name *infra-red* is derived from the Latin word *infra*, meaning below, and the English word *red*. As the name infra-red implies, it simply means below red; therefore, we cannot see it. It is due to this aspect of infra-red that the military uses infra-red to "spot" enemy troop movements at night.

No practical application of Sir Herschel's isolation and determination of infra-red was found for using infra-red for building heating until early in

World War II. It was then in the late 1930s and early 1940s that it was found. It was rumored that during World War II, Germany had difficulty heating its large manufacturing facilities that were producing critical war materials for its military. Supposedly, Adolf Hitler charged a German scientist, Dr. Herman Schwank, to find a more effective and less expensive way to heat these large facilities so as to increase production of these critically needed war materials. Apparently, Dr. Schwank was impressed with Sir Herschel's discovery and experimented with ways to generate infra-red and use it to heat these buildings. He then devised and designed a gas-fired perforated ceramic burner that created a high-temperature infra-red surface using, at the time, manufactured gas. Now he had the means to create a gas-fired infra-red heater. As in the case with most rumors, this rumor proved to be false. Actually Dr. Schwank experimented with gas infra-red burners during the late 1920s. He was manufacturing and selling his early units as early as 1933. His early units were sold and installed in certain areas of Germany and France. Thereby came the introduction of the gas-fired infra-red heater that we know today.

Prior to 1955, most people associated infra-red with electricity and with high operational costs. Because of cost limitations, electric infra-red was and still is limited to spot and area heating and for certain process applications.

With the eventual introduction of natural gas, gas-fired infra-red heaters were used for building heating. This unique new method of gas heating grew rapidly. This was especially true with total building heating in industrial and commercial applications. Other unusual uses include process heating applications, snow melting, and corrosion prevention.

This book deals basically only with gas infra-red as a means to heat entire buildings and will include basic theories and application highlights exclusively. If greater details are required, it will be necessary to refer to Heat Transfer textbooks and manufacturer's application manuals. The infra-red heaters will be the unvented variety and conform to ASA Z21.43, Approval Requirements for Unvented Gas-Fired Infra-Red radiant heaters.

Competitive forms of conventional heaters will be dealt with for comparison purposes. There are three basic methods available to transfer heat from a heater to a cooler area: they are convection, conduction, and radiation. Unit heaters rely basically on conduction, and convection or a combination of the two to achieve comfort heating.

Conduction occurs as the result of faster moving molecules (higher temperature) that transfer some of its energy to adjacent molecules that are at a lower temperature. This phenomenon may occur within a solid or between a solid and an adjacent fluid such as water or air.

Convection of heat is the result of the mixing of warm and cool particles of density differences due to temperature differences, which is natural convection. Should the mixing be produced by mechanical means, it is forced convection.

Radiation heat transfer is different from conduction or convection in that it does not require matter to accomplish the transfer. Radiated heat is usually known as infra-red.

Introduction to IR Heating

As a young sales engineer working for a gas utility during the early 1960s, I was introduced to the IR comfort heating concept by several IR heater manufacturers' representatives, namely Ed McIntosh, Augie Carell, John Purkey, and Milt Wolf. Later, in my search for a more specific IR design, I was privileged to meet, in my opinion, two of the truly great minds of gas-fired IR heating design and applications–Fred Prince and Bill Best.

It was through these people that I learned how to properly design, sell, and appreciate the advantages of IR comfort heating systems. The economic advantages and the energy-saving aspect of gas IR systems make the design and sale a definite win-win application.

This book is written based on proven design, sales, and installation experience with gas-fired IR systems. The purpose of this book is to show anyone interested in the design, economic advantages, as well as the energy conservation advantages of gas-fired IR heating systems.

As is the case with any comfort heating equipment or systems, there are disadvantages as well as advantages that need to be addressed. These positives and negatives will be discussed in greater detail later in the text, to avoid what could be serious design problems. Hopefully, the design and application procedures described in the text can be used: by the reader, in informal self-study courses, in-house training programs or as a supplemental text in college or technical school courses on energy-efficient HVAC systems.

Included later in this text are some of the varied additional process applications and uses of gas-fired infra-red energy. Industrial applications such as paint baking in the auto industry, heat treating of metals, curing of masonry, drying of textiles and paper, brick and tile manufacturing, etc., are just some of the uses for gas-fired infra-red. The food industry finds gas-fired IR ideal for roasting meats and broiling steaks because infra-red burners penetrate deep, cook faster, reduce shrinkage, and thereby improve the flavor and taste of cooked meat.

List of Figures

List of Tables

List of Abbreviations

ACGIH	American conference of governmental industrial hygienists
AGA	American gas association
ASHRAE	The American society of heating, refrigerating and Air-Conditioning engineers
BIP	Black iron pipe
CO	Carbon monoxide
DB	Dry bulb
DD	Degree days
ERV	Energy recovery ventilator
FLA	Full load amps
FM	Factory mutual
FPM	Feet per minute
HVAC	Heating, ventilation, and air conditioning
IAQ	Indoor air quality
IR	Infra-red
OD	Outside diameter
PE	Professional engineer
PPM	Parts per million
RCSA	Space Ray's designation for its high-intensity IR heaters as opposed to its SRO-40 units
RH	Relative humidity
ROR	Rate of return
TEC	Thermal engineering corp

Introduction

Basic Heating

Heating of any building that has a heat loss requires some form of heating equipment with some form of fuel to supply the heat source. Prior to and even after Colonial times wood, in the form of logs, was used as fuel in the fireplaces. Wood was plentiful with vast forests available that were cleared for farming and also used for fuel. Storage of logs was always a problem, as well as splitting the logs for better combustion. Abraham Lincoln is known for this to earn a living. Burning wood in fireplaces caused many house fires in those days. Fortunately there was the "pot belly" stove that gradually replaced the fireplace to burn wood and made building heating safer. Benjamin Franklin came up with an improved design, known as the Franklin stove. The Franklin stove was more efficient than the pot belly and definitely better than the fireplace.

Coal

When coal became a readily available fuel, it was used to replace wood in the fireplace as well as the pot belly and Franklin stoves. The advantages were: less storage space was required, it was a slower burning fuel, easier to control, safer, and cleaner.

The Franklin stove is still in use in some sections of the country for winter vacation homes and some year round homes where oil and gas are not available. The Franklin stove burns wood or coal to produce convected heat into the room to heat the air, but it also provides radiant heat in the form of infrared into the space. Anyone in the room can feel this radiant heat being emitted from the sides of the stove. Pot belly stoves operate in the same manner.

Boilers and Furnaces

With coal now available, heating equipment designers began to design and manufacture boilers that could generate and supply steam to heat multiple

DOI: 10.1201/9788770046589-1

spaces/rooms and for all types of heating applications using steam piping and radiators. Commercial and industrial requirements as well as comfort heating could now be possible.

In addition to boilers, manufacturers also designed and manufactured coal-fired furnaces. Here air is heated through a heat exchanger into ducts that conduct the heated air into the spaces requiring comfort heating.

The storage space for coal is not the only problem that must be addressed. The removal of ashes is always a problem, along with labor to stoke and shovel the coal.

Fuel Oil

Eventually, John D. Rockefeller and other oil producers provided a solution to the replacement of coal–fuel oil. Fuel oil was easier to use in boiler plants, and it was easier to store than coal and wood. Oil also reduced the labor costs as well as fuel costs. It became obvious that fuel oil would replace coal as the premium heating medium. Unit heaters utilizing steam and hot water became a popular way of providing comfort heating. However, steam unit heaters require supply piping to get the steam to the units and piping to return the condensate to the boiler, plus steam traps in the supply lines to control the flow of steam. Hot water unit heaters have basically the same installation problems. Piping is needed to supply the hot water and piping for the return water to the boiler. In addition a pump is required to provide and circulate the hot water for comfort heating.

In order to reduce these installed costs, direct oil-fired unit heaters began to replace steam and hot water systems in initial comfort heating design.

Natural Gas

Next in line of available fuels was manufactured gas, which was eventually replaced with natural gas. Natural gas solved: the storage problem, the oil piping problem, and the fuel pump problem. Natural gas, where available, is provided from gas mains in the street, through a gas service into the building (installed by the gas utility) through a gas meter into the building. The gas is paid for after usage. The one drawback is the cost of the fuel on a Btu basis. It gets a little worse with #4 oil at 150,000 Btu/gal and with #6 oil, which is basically tar at 160,000 Btu/gal. However, #6 oil must be heated in order for it to flow, and in some cases #4 oil needs to be heated in order to burn efficiently.

This is where gas-fired infra-red heaters come into the competition with oil. None of the three grades of oil can compete with gas-fired infra-red building heating.

Infra-red energy is as old as the sun. However, the generation of infra-red energy using natural gas is relatively new. Basically, any object that is warmer than surrounding objects or materials is an infra-red generator. For this reason, a human being could be considered an infra-red generator.

Actually radiant heating equipment has demonstrated its practical nature in many diverse applications. Literally, there are thousands of cost-effective installations that have been installed in small machine shops, as well as large high bay structures such as airplane hangars, steel warehouses, and gymnasiums.

It is important in the design process to focus on unusual factors in the building's usage. The designer should determine if there are potential hazardous or corrosive vapors that exist. Whether there are combustible materials or products that are being stored up to the ceiling or forklifts that will be raising merchandise high onto storage shelves. Sometimes the mounting height of the IR heaters will be determined by the mounting of the lighting fixtures in the area. Another consideration that should be addressed is how close the units will be to the underside of the roof. It may be necessary to install heat shields above the heaters.

Hazardous Conditions

It may be necessary to check the local fire codes to help decide whether infra-red is eligible for building heating. Should there be vapors present, it may make sense to design a fresh air supply to create a self-dependent system that does not use any inside building air for ventilation.

Corrosive Environment

There will be times when products such as cleaners, preservatives, chemicals, or manufacturing procedures will introduce corrosive or hazardous conditions. This is especially true with products containing chlorine or certain chlorides.

Electric Infra-Red

Unfortunately when you talk to most people about infra-red heating, they automatically think of electric infra-red. However, electric infra-red cannot realistically be used for total building heating–only spot heating at best. As an example: to duplicate a small 30,000 BTUH gas, the infra-red heater would require an electric infra-red unit utilizing 8.79 kW, which would draw 36 A at 240 V or 73 A at 120 V. If the building needed just ten of the 30,000 BTUH units, which is typical, it would literally require a small substation to provide power just for heating. With a three-phase 120/208 volt service, it would need a 300 A service just for the heating load. Add to the cost of the 300 A additional service capacity the cost of internal wiring that is required for each 8.79 kW infra-red heater. We are looking at major costs with switch gear requirements and labor costs. So much for electric infra-red building heating!

Gas-Fired Infra-Red

In order to effectively generate IR, it requires efficient conversion from electric or gas energy to infra-red radiation. Methods for accomplishing this are: utilization of special alloy screens; properly designed ceramic emitters; special burners for providing infra-red with tubes and other gas-fired equipment that generate the desirable infra-red with maximum efficiency. Infra-red rays are directional and can be focused like a flashlight.

Engineers and Architects

Gas-fired infra-red (IR) comfort heating is the least understood and probably the most unappreciated form of heating system in this country–or even in the world. This is possibly due to engineers and architects being unaware of the economic advantages of IR heating as opposed to their old central boiler plant designs City regulations that have not been upgraded to the newer and more efficient equipment and designs may also be a factor in their thinking. Schools, gymnasiums, churches, entertainment centers, and many public buildings are all excellent examples of archaic designs and regulations. Most engineers and architects designing school heating specify a central boiler plant that includes the gymnasium and auditorium along with the classrooms and offices, in the sizing of the boiler.

Schools with Gymnasiums

As an example, schools with gymnasiums present a typical HVAC engineer with many design problems. Schools and hospitals are required by law, in many states, to have a "backup" boiler of the same boiler capacity. This is to be certain that there is heating capacity available should the main boiler fail, for any reason, during the heating season. The gymnasium adds to this problem. With its high ceiling, large floor area, and ASHRAE ventilation requirements, the large heat loss adds to the central boiler size. This amounts to many additional boiler horsepower onto the central boiler plant, including the backup unit. Economically, this is "overkill".

Unusual Operating Conditions

Since basketball season coincides with the heating season and the boiler has been sized to heat not only the gym but the classrooms, auditorium, and offices, the large boiler is used just to heat the gymnasium. This occurs for the simple reason that most games are played at night. A simple solution to this design problem would be gas-fired infra-red heating for the gymnasium and modular boilers for the basic heating load. The same can be said for auditoriums. This can be accomplished using low-temperature infra-red tube heaters. They would be unobtrusive and could easily blend in with the auditorium décor. In addition, the low-intensity units would be quieter than any fan coil blowers used to provide heating from the central boiler plant.

It is a known fact that heating boilers achieve rated efficiency only when operated at full-rated input for extended periods of time. Based on this fact, large capacity boilers operate intermittently when the load is reduced and operate in short inefficient combustion cycles. The same can be said for an auditorium. Short-cycling in effect reduces boiler efficiency. Depending on the percentage of "on-time," it is estimated that the boiler efficiency can range from 66% efficiency at 20% on-time to 75% when it is used at continuous operation.

Modular Boilers with Gas Infra-Red

This condition can be rectified by installing infra-red heating in the gymnasium and/or the auditorium and replacing the over-sized central boiler plant with a modular boiler system. This could also overcome many city codes requiring a full-sized back-up boiler. Experience has shown that for 90% of the entire heating season, only 60% or less of the boiler capacity is required. With modular boilers, this can be easily be met even with one or two of the modular units inoperable.

With gas infra-red in the gymnasium, it will also reduce noise levels and keep the floor warm in the gyms. Annual energy costs could easily be reduced by at least 30% and possibly as much as 50%.

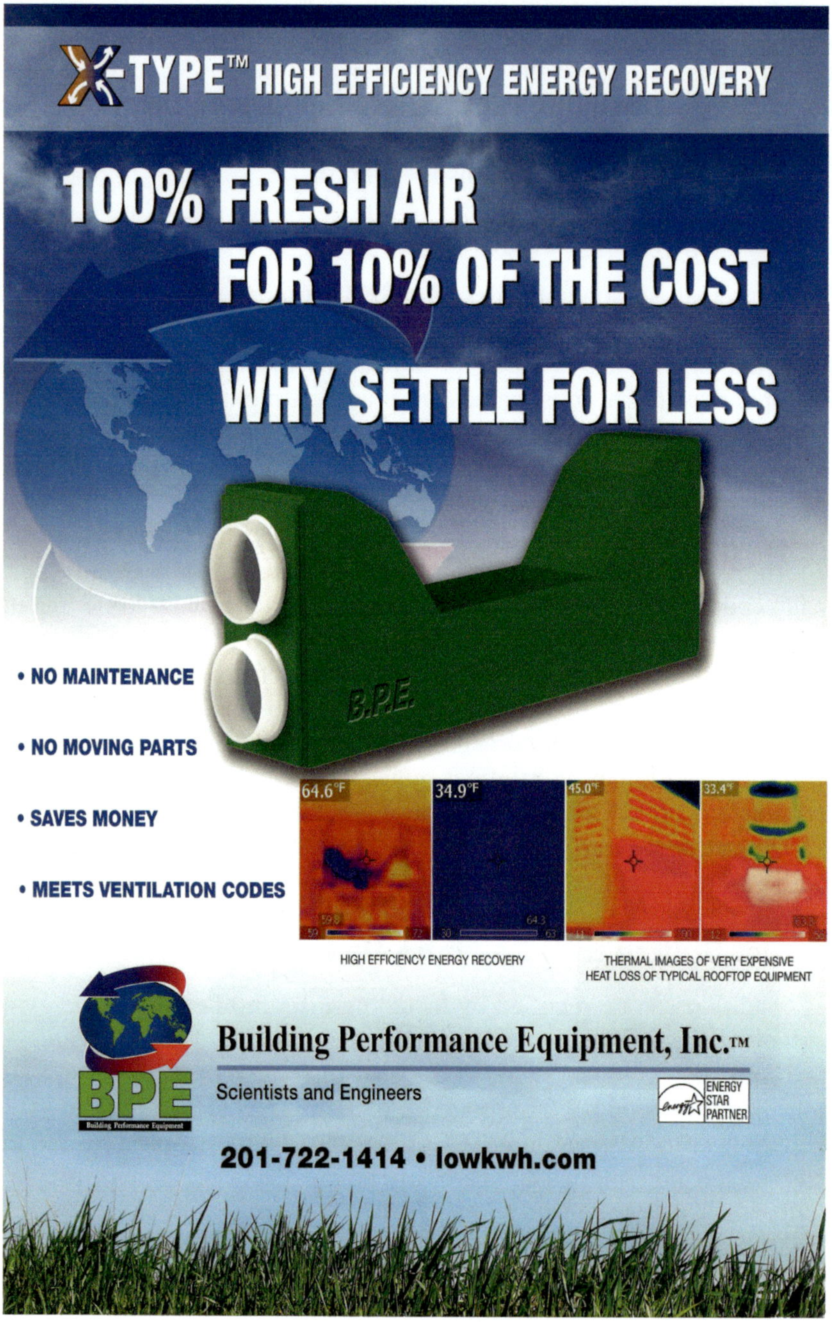
X-TYPE™ HIGH EFFICIENCY ENERGY RECOVERY
100% FRESH AIR
FOR 10% OF THE COST
WHY SETTLE FOR LESS
B.P.E.
• NO MAINTENANCE
• NO MOVING PARTS
• SAVES MONEY
• MEETS VENTILATION CODES
64.6°F
34.9°F
45.0°F
33.4°F
HIGH EFFICIENCY ENERGY RECOVERY
THERMAL IMAGES OF VERY EXPENSIVE
HEAT LOSS OF TYPICAL ROOFTOP EQUIPMENT
BPE
Building Performance Equipment
Building Performance Equipment, Inc.™
Scientists and Engineers
ENERGY STAR PARTNER
201-722-1414 • lowkwh.com

The Latent Side of BPE
Eco Air Anywhere®

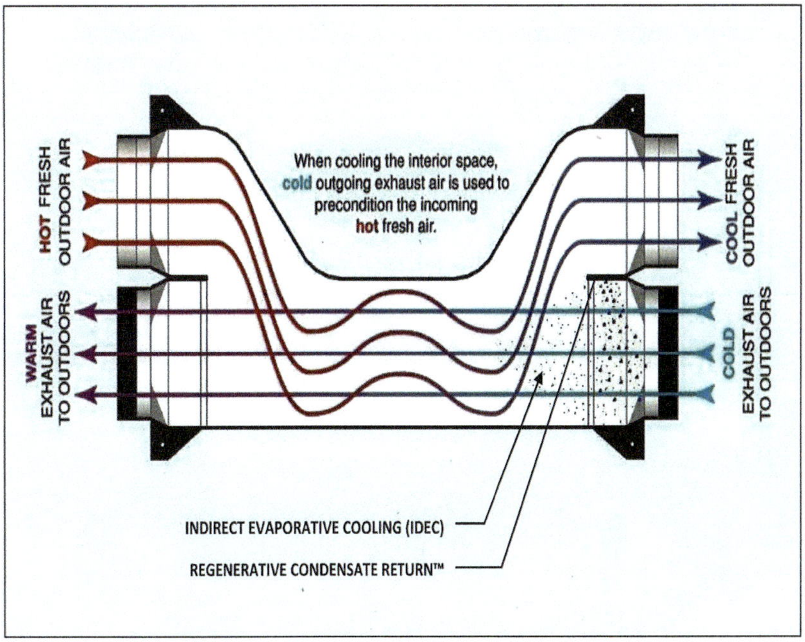

BPE's Patented Regenerative Condensate Return™ Technology

While traveling through the core of the ERV, hot and humid fresh air will be cooled within a few degrees of room temperature. As the humid air cools, it will produce excess condensation. This condensation must be removed before the air can be used. Water can create rust and corrosion which can decrease equipment lifetimes and increase maintenance costs. Using BPE's patented ***Regenerative Condensate Return® (RCR)*** Technology, excess condensation is transferred into the exhaust stream rather than into your building.

One huge benefit of the RCR Technology is not only its ability to create a latent effectiveness, but its ability to increase the sensible effectiveness. As condensation is absorbed by the dry exhaust air, it will begin to evaporate. This effect is created by the heat exchanged from the hot outdoor air to the exhaust air. Evaporation of the moisture creates a cooling effect and increases the sensible effectiveness of the heat exchanger core. This process is called Indirect Evaporative Cooling (IDEC), and no one does IDEC better than BPE.

Building Performance Equipment, Inc.™

Scientists and Engineers

X-TYPE UNITS ARE HIGH EFFICIENCY (OVER 85%) AND CAN SUPPLY FRESH PRECONDITIONED AIR TO A COMMON HALLWAY **COMFORTABLY**.

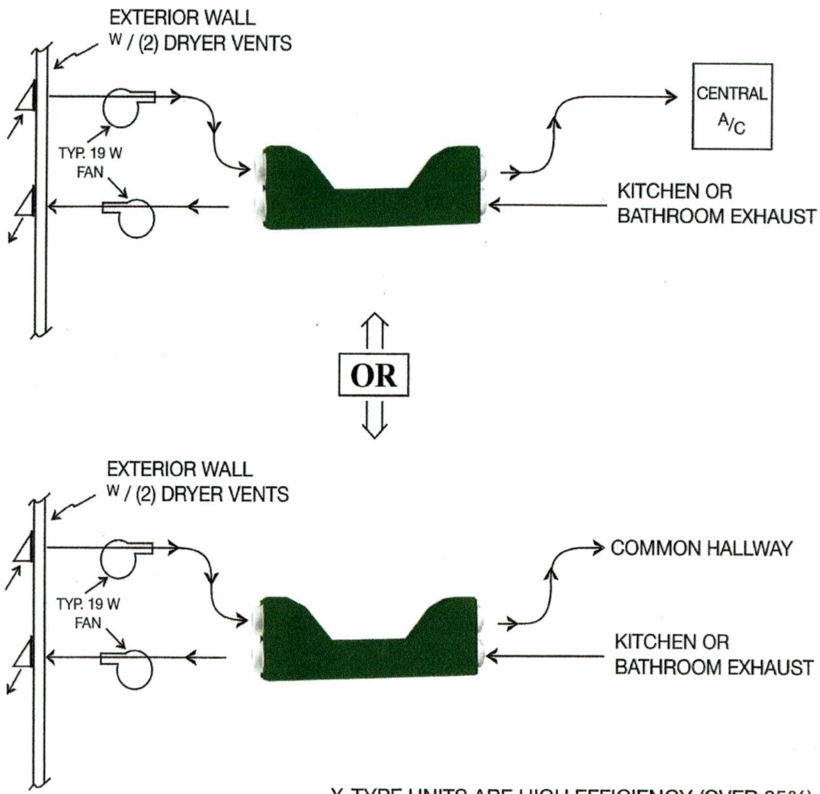

X-TYPE UNITS ARE HIGH EFFICIENCY (OVER 85%) AND CAN SUPPLY FRESH PRECONDITIONED AIR TO A COMMON HALLWAY **COMFORTABLY**.

PATENTS AND PATENTS PENDING

80 Broadway, Hillsdale, New Jersey 07642
Voice: (201) 722-1414 • FAX: (201) 722-0999 • lowkwh.com

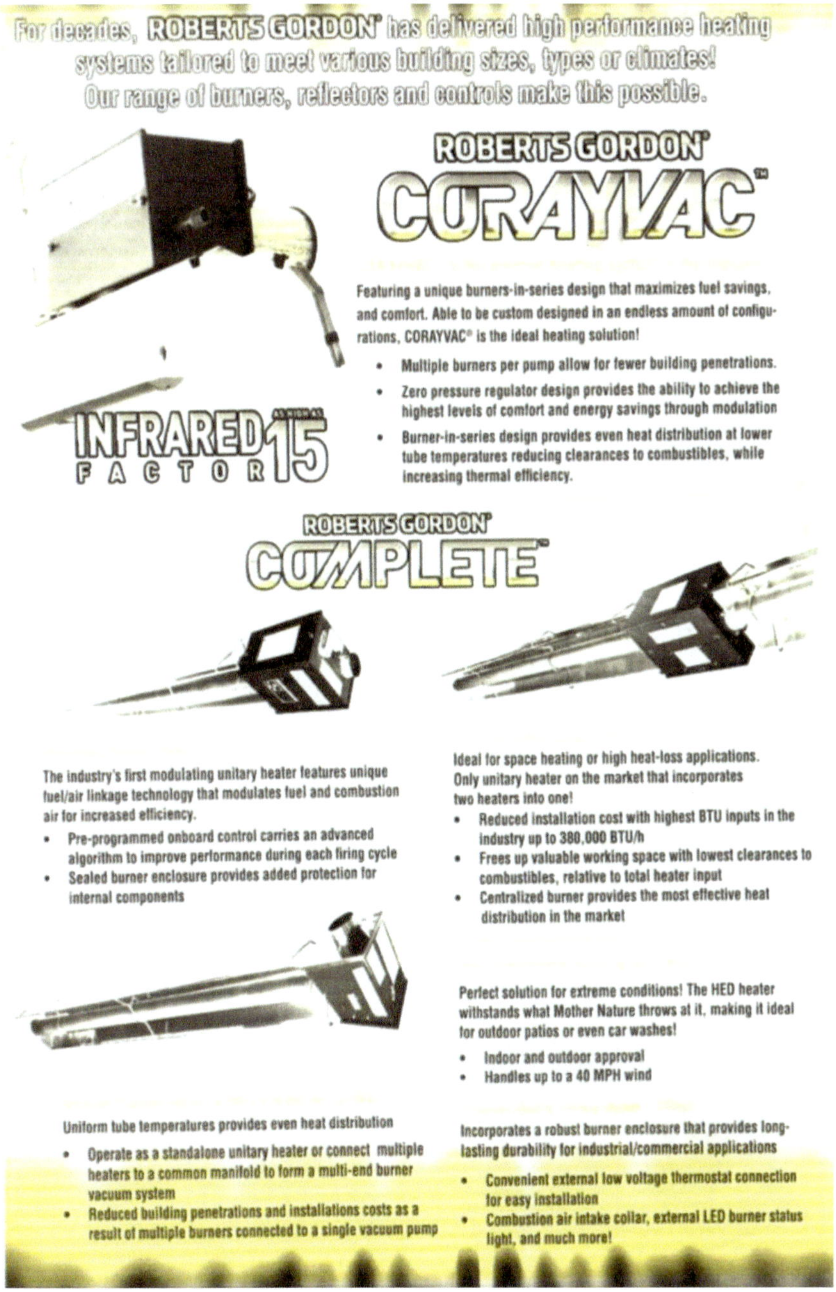
For decades, ROBERTS GORDON® has delivered high performance heating systems tailored to meet various building sizes, types or climates!
Our range of burners, reflectors and controls make this possible.
ROBERTS GORDON®
CORAYVAC™
Featuring a unique burners-in-series design that maximizes fuel savings, and comfort. Able to be custom designed in an endless amount of configurations, CORAYVAC® is the ideal heating solution!
• Multiple burners per pump allow for fewer building penetrations.
• Zero pressure regulator design provides the ability to achieve the highest levels of comfort and energy savings through modulation
• Burner-in-series design provides even heat distribution at lower tube temperatures reducing clearances to combustibles, while increasing thermal efficiency.
INFRARED FACTOR AS HIGH AS 15
ROBERTS GORDON®
COMPLETE™
The industry's first modulating unitary heater features unique fuel/air linkage technology that modulates fuel and combustion air for increased efficiency.
• Pre-programmed onboard control carries an advanced algorithm to improve performance during each firing cycle
• Sealed burner enclosure provides added protection for internal components
Ideal for space heating or high heat-loss applications. Only unitary heater on the market that incorporates two heaters into one!
• Reduced installation cost with highest BTU inputs in the industry up to 380,000 BTU/h
• Frees up valuable working space with lowest clearances to combustibles, relative to total heater input
• Centralized burner provides the most effective heat distribution in the market
Perfect solution for extreme conditions! The HED heater withstands what Mother Nature throws at it, making it ideal for outdoor patios or even car washes!
• Indoor and outdoor approval
• Handles up to a 40 MPH wind
Uniform tube temperatures provides even heat distribution
• Operate as a standalone unitary heater or connect multiple heaters to a common manifold to form a multi-end burner vacuum system
• Reduced building penetrations and installations costs as a result of multiple burners connected to a single vacuum pump
Incorporates a robust burner enclosure that provides long-lasting durability for industrial/commercial applications
• Convenient external low voltage thermostat connection for easy installation
• Combustion air intake collar, external LED burner status light, and much more!

CORAYVAC® High Efficiency, Condensing Infrared System Features

REFLECTORS

ROBERTS GORDON® deep-dish, standard aluminum reflectors are shaped to help direct reflection of the energy emitted by the infrared tube and beam towards the floor where needed. Our industry-leading high efficiency reflectors have a wider design that maximizes the amount of radiant energy emitted from the heat exchanger. This directs an even greater amount of radiant energy towards the floor which results in increased energy savings.

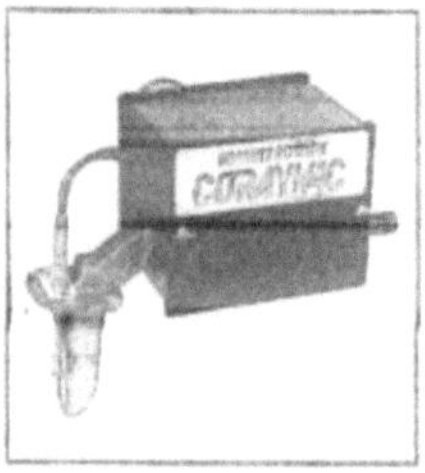

BURNERS-IN-SERIES DESIGN

Located from 10 ft. - 70 ft. (3 m to 21.3 m) apart CORAYVAC® burners constantly regulate the air-to-gas mixture to achieve the optimum ratio for clean, efficient combustion. Models are available with inputs of 20,000 through 120,000 Btu/h. Features include three-try direct spark ignition, pre-purge, filtered combustion air and cast-iron burner head.

Uniform Heat with CORAYVAC® In-Series Burners

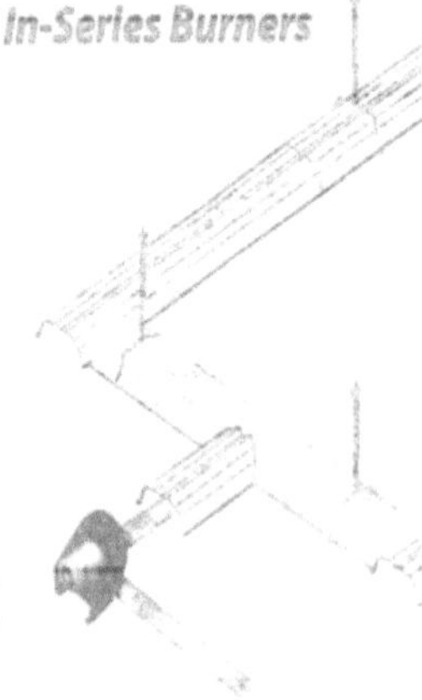

TUBING

The CORAYVAC® system utilizes 4" O.D. (100 mm), 16-gauge tubing. The heat created by the burners is drawn through the tubes, which radiate the warm, gentle, infrared energy. Hot-rolled steel, aluminized steel or double porcelain coated steel tubing are available. Double porcelain coated steel is a cured porcelain coating on the inside and outside surface of the tube, which helps to maximize longevity and minimize corrosion of condensing systems.

VACUUM PUMPS

A vacuum pump draws the heat throughout the entire system. It exhausts products of combustion to the outdoors at temperatures typically below 150° F (66° C). 1/3-hp, 3/4-hp or 2-hp are available for various system sizes and layouts. Up to twelve heaters can be common vented with one vacuum pump. Design flexibility allows side wall venting, even in large buildings.

Schwank
supraSchwank
Infrared heating with energy savings up to 65%
supraSchwank overhead luminous gas heaters can be mounted on ceilings up to 180 feet high and will still provide superior comfort and temperature levels for people below with up to 65% energy savings!
Perfect for industrial and commercial spaces and aircraft hangers, these radiant heaters have been awarded with the highest efficiency in North America. supraSchwank offers the highest conversion rate of any IR heater in the market at 81%*
Schwank is the worldwide market leader and most trusted brand with over 2 million heating appliances sold.
Features
81% radiant Efficiency*
Patented Delta Mixing Chamber preheats gas/air combustion mixture for uniform and ultra efficient heat delivery to work level
Fully insulated heater made of Feran to minimize convection heat losses
Green technology: fuel costs and CO2 emissions reduced up to 65%*
Optional Accessories
ThermoControl Plus
TruTemp
Lion Chain
1-877-446-3727 | car@schwankgroup.com | www.schwankgroup.com

K-SERIES
HEATERS

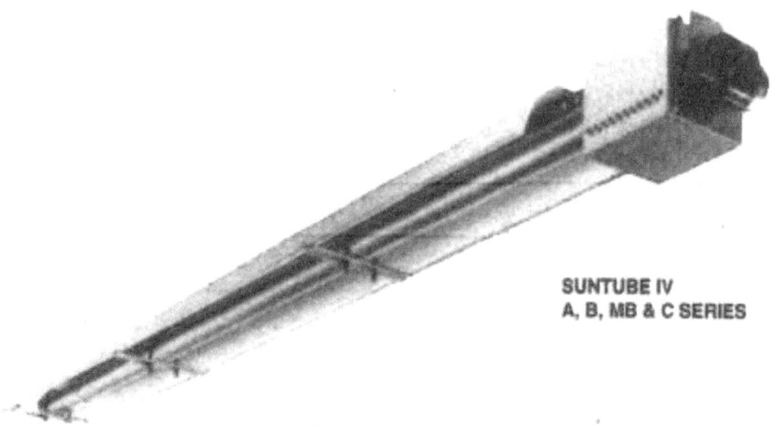

SUNTUBE IV
A, B, MB & C SERIES

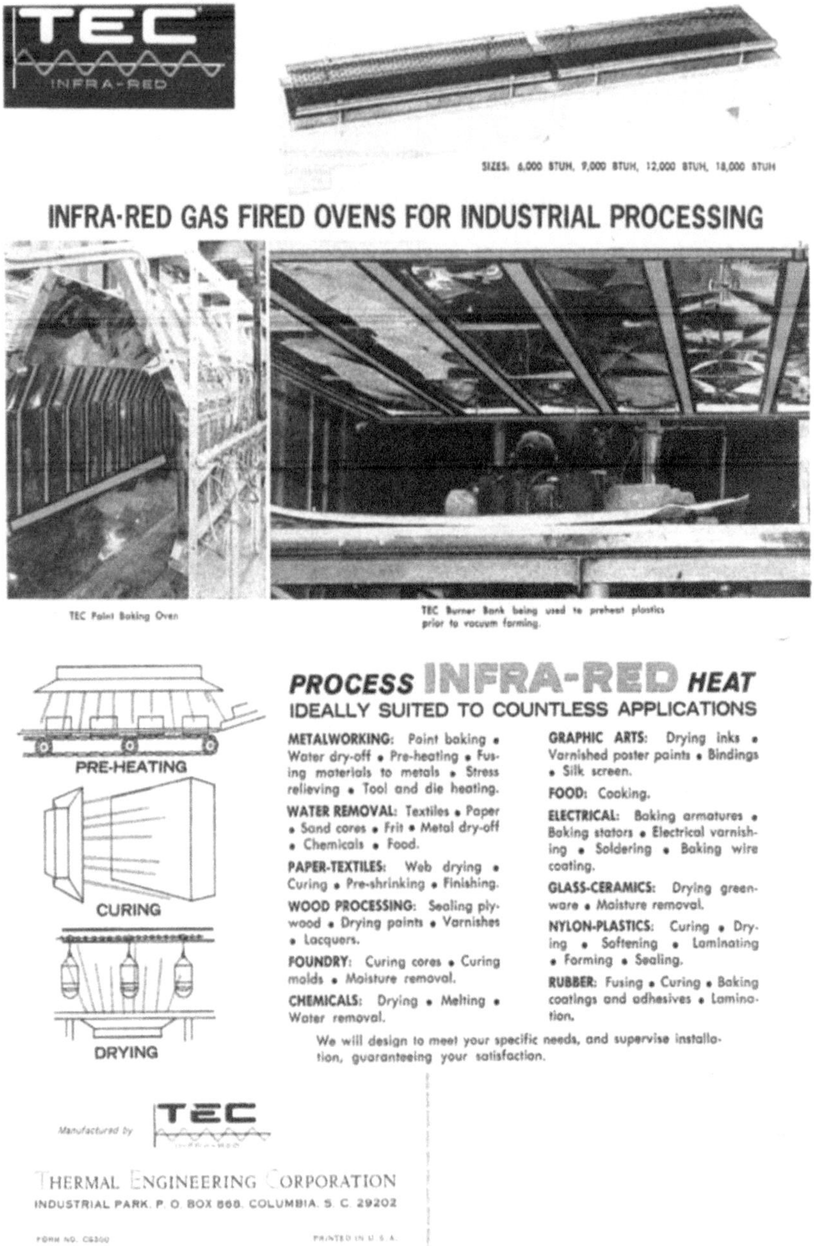
TEC
INFRA-RED
SIZES: 6,000 BTUH, 9,000 BTUH, 12,000 BTUH, 18,000 BTUH
INFRA-RED GAS FIRED OVENS FOR INDUSTRIAL PROCESSING
TEC Paint Baking Oven
TEC Burner Bank being used to preheat plastics prior to vacuum forming.
PRE-HEATING
CURING
DRYING
PROCESS INFRA-RED HEAT
IDEALLY SUITED TO COUNTLESS APPLICATIONS
METALWORKING: Paint baking • Water dry-off • Pre-heating • Fusing materials to metals • Stress relieving • Tool and die heating.
WATER REMOVAL: Textiles • Paper • Sand cores • Frit • Metal dry-off • Chemicals • Food.
PAPER-TEXTILES: Web drying • Curing • Pre-shrinking • Finishing.
WOOD PROCESSING: Sealing plywood • Drying paints • Varnishes • Lacquers.
FOUNDRY: Curing cores • Curing molds • Moisture removal.
CHEMICALS: Drying • Melting • Water removal.
GRAPHIC ARTS: Drying inks • Varnished poster paints • Bindings • Silk screen.
FOOD: Cooking.
ELECTRICAL: Baking armatures • Baking stators • Electrical varnishing • Soldering • Baking wire coating.
GLASS-CERAMICS: Drying greenware • Moisture removal.
NYLON-PLASTICS: Curing • Drying • Softening • Laminating • Forming • Sealing.
RUBBER: Fusing • Curing • Baking coatings and adhesives • Lamination.
We will design to meet your specific needs, and supervise installation, guaranteeing your satisfaction.
Manufactured by
TEC
THERMAL ENGINEERING CORPORATION
INDUSTRIAL PARK, P. O. BOX 868, COLUMBIA, S. C. 29202
PRINTED IN U.S.A.

Part I

DOI: 10.1201/9788770046589-2

1

Theory of Thermal Radiation

Every object at temperatures above absolute zero gives off infra-red energy. The amount of infra-red given off is a direct function of the temperature, area and the emissivity or radiant efficiency of the object. When the object is heated to the point it glows, it gives off visible light. If we heat two objects of identical material of different sizes, heated to the same temperature, the larger object will radiate more heat than the smaller one. However, if we heat the smaller object to a higher temperature, we can make it radiate the same amount of infra-red as the larger one. We must be careful not to heat the small object too much because it will become incandescent and begin to emit light energy. Instead of being absorbed as usable heat, it may be reflected. All solid objects absorb radiant energy to varying degrees depending on the surface texture, composition, color, and the wave length of the radiant energy meeting the surface. The wave length again is the function of the source temperature. The higher the temperature, the shorter the wave length. Infra-red energy, when striking a surface, generates heat that transfers heat through the surface being heated by conduction.

Heat may be transmitted from one body to another without changing the temperature of the air in between. This can be illustrated with a very simple experiment. When standing before a fireplace, you can feel the warmth that is not due to the temperature of the air, because if a solid screen is set in place between you and the fireplace, the feeling of warmth disappears. However, if a glass screen or a transparent screen is used, you would still be warmed by the infra-red emitted from the fireplace. Infra-red, unlike ultra-violet radiation, does transmit through glass. Proof of this fact is you can feel the infra-red heat from the sun, but you cannot get a sunburn or suntan from infra-red. Ultra-violet provides the "tanning" rays from the sun, but not through glass.

This phenomenon, known as thermal radiation, is just one of the many forms of radiant energy continuously being released and absorbed, in various

DOI: 10.1201/9788770046589-3

levels by all bodies. All radiant energy should be recognized as a form of wave motion known as "electromagnetic phenomena." Electromagnetic waves should not be confused with sound waves and other mechanical waves that only occur in solids, liquids, and gases; whereas, radiant energy waves can be transmitted even through a vacuum. Thermal radiation is radiant energy emitted because an object has heat. It is sometimes called infrared heat emittance, radiant flux density, or emissive power. Its measurement is in Btu per square foot per hour. As a result of this interchange of radiation and absorption, there is a continuous reciprocation of energy between all bodies. Therefore, if separate bodies are at different temperatures within a building, the warmer body receives less energy from the cooler one than it radiates, and as a result its temperature decreases. Consequently, the cooler body receives more heat than it gives off and its temperature increases.

This exchange of heat continues even after each body reaches the same temperature, with each body receiving as much heat as they radiate. Based on this concept, a body ceases to emit thermal radiation only when it reaches the absolute zero temperatures or $-460\,^{\circ}$F.

The dictates of radiation show that the higher the temperature of the emitting body, the faster it gives off infra-red energy. From this, it can be seen that the amount radiated per unit of time will be proportional to the surface area of the emitting body.

Heat always transfers from hot to cold. Therefore, if we install an infra-red heater above a cold concrete floor, the infra-red heater will transmit heat energy to that cold floor until it becomes warm. Any machinery, equipment, or objects at floor level will also become warm from direct radiation as well as conduction and convection from the warm floor slab. The floor slab then becomes a giant heat sink, radiating to the sidewalls and roof structure. Infra-red is the only form of heat that can effectively heat a floor economically.

Infra-red radiation is an energy that is produced as an electromagnetic wave similar to radio waves, light, and X-rays. These waves travel at the speed of light (186,272 miles per second). While being emitted in this manner, the energy is called "radiant energy" and is not heat. This energy becomes heat instantly when it is absorbed by an absorbing material.

Absorptivity is the fraction of thermal radiation that is absorbed by an object or surface. Any given body under a fixed set of conditions will emit radiation of various wave lengths. The amount of intensity of the radiation in the various wave lengths is different. The type of radiation emitted from a certain body is characterized by the band of wave lengths having the greatest intensity. The curves of Figure 1.1 show the distribution of intensity of

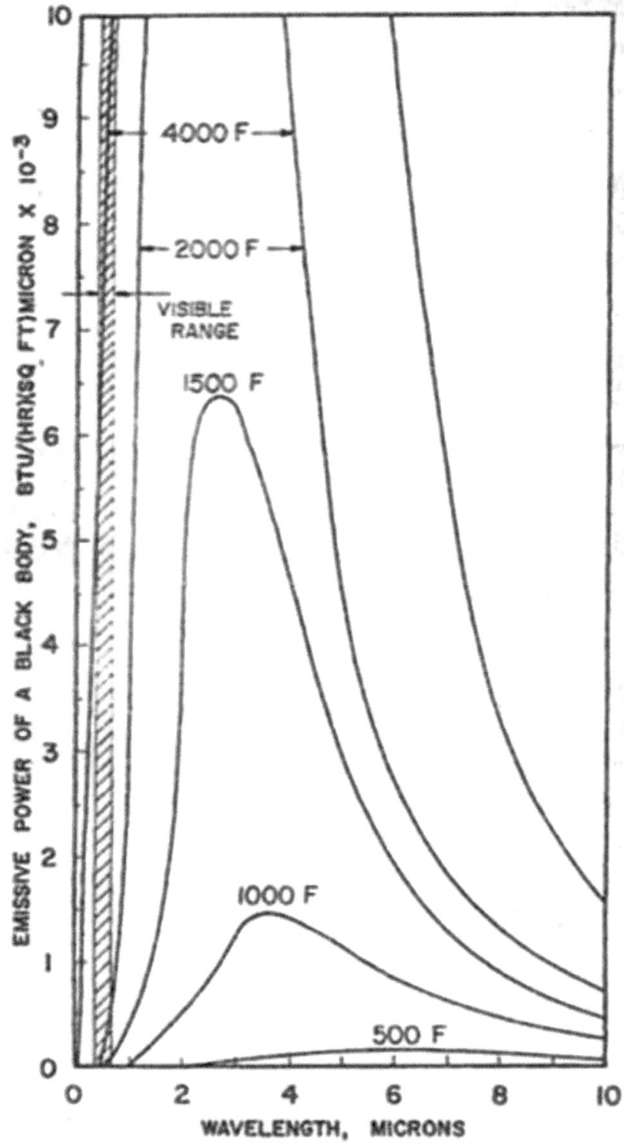

Figure 1.1 The Emissive Power or radiant energy generated at any wave length by a "black body" – the most efficient surface emitting thermal radiation – is shown here as a function of the temperature of the surface. As the temperature of the surface increases, the emissive power increases and the wave length at which the maximum radiation occurs decreases. The total radiation emitted at any given temperature is equal to the area under the curve for the temperature.

radiation with wave lengths for a "black body." A "black body" is one that emits the maximum possible radiation at a given temperature. The adjective black has nothing to do with the color of the emitter. Table 1.1 provides the same information; however, it is in tabular form. An examination of Table 1.1 or the curves of Figure 1.1 would result in the following conclusions:

1. Any increase in temperature causes a decrease in the wave length at which maximum energy emission occurs.
2. Any increase in temperature causes a rapid increase in energy emission at any given wave length and in the total energy emission.
3. The total rate of energy emission at any temperature and for any range of wave lengths is given by the area under the curve for that temperature.

Absorptivity of a "black body" equals 1.0 and, therefore, its reflective ability equals zero. Although no totally black surface exists in nature, the concept of a "black body" is a useful one because the laws governing its radiation are comparatively simple and because it can be proved that no surface radiates more heat than a "black one" [1].

Infra-red energy generated by a "blackbody" depends on its absolute temperature only and can be computed by use of the Stefan–Boltzmann law, which states:

$$E_{\mathrm{bb}} = \sigma T^4,$$

where:

E_{bb} = total thermal radiation emitted by a "blackbody"

σ = Stefan–Boltzmann constant $= 0.1714 \times 10^{-8}\,\mathrm{BTUH/(ft^2\,{}^\circ R^4)}$

T = Absolute temperature °R (Rankine)

The above "black body" law and equation has been based on having a perfect radiator. Since infra-red devices are not perfect, modifications of the "black body" laws are needed to provide good engineering estimates of thermal radiation emitted by the IR heater. By including the emissivity of the actual emitter and its effective area in the Stefan–Boltzmann law is a good approximation of the thermal radiation that can be made. The formula then becomes

$$E_{\mathrm{ACT}} = A\varepsilon\sigma T^4,$$

Table 1.1 Black Body Thermal Radiation – BTU/HR-FT^2

(1) IN VARIOUS WAVE LENGTH BANDS (QUALITY)
(2) TOTAL THERMAL RADIATION (QUANTITY)
(3) AND WAVE LENGTH AT MAXIMUM RADIATION

Source Temp °F	(1) WAVE LENGTH BAND - MICRONS										(2) Total Thermal Rad.	(3) Wave Length At Max. Rad. Microns
	0–0.7	0.7–1	1 - 2	2 - 3	3 - 4	4 - 5	5 - 6	6 - 8	8 - 10	10 +		
500			1	27	99	155	174	308	215	482	1460	5.43
600			3	70	201	278	285	442	300	590	2170	4.92
700			12	162	376	445	414	620	376	705	3110	4.50
800			32	325	620	672	575	809	468	840	4340	4.14
900			76	576	963	952	752	1035	565	960	5880	3.84
1000			174	960	1400	1305	975	1250	660	1095	7820	3.57
1100			316	1525	1982	1628	1245	1510	765	1220	10,190	3.34
1200		1	556	2250	2673	2053	1518	1763	876	1370	13,060	3.14
1300		3	973	3200	3424	2550	1830	2010	1025	1483	16,500	2.96
1400		8	1564	4404	4380	3090	2059	2340	1095	1650	20,590	2.80
1500		18	2388	5960	5335	3600	2415	2640	1245	1780	25,380	2.66
1600		34	3557	7720	6410	4340	2715	2910	1365	1920	30,970	2.53
1700		71	5059	9840	7570	5050	3035	3215	1540	2060	37,440	2.41
1800		126	7040	21,310		5525	3450	3590	1620	2200	44,870	2.31
1900	5	225	9550	25,440		6300	3840	3890	1760	2350	53,360	2.21
2000	12	13,090		29,935		11,210		4220	1965	2455	62,990	2.12
2500	157	43,125		58,650		17,300		6015	2500	3155	131,200	1.76
3000	1281	109,660		96,850		23,900		7640	3205	3940	246,400	1.51
4000	20,415	405,560		194,620		38,110			21,780		680,500	1.17
5000	129,900	1,005,400		311,750		51,950			29,030		1,528,000	0.96

THERMAL ENGINEERING CORP.
COLUMBIA, SOUTH CAROLINA

where:

E_{ACT} = Actual thermal radiation

A = Effective or solid area of the emitter

ε = Emissivity

σ = Stefan–Boltzmann constant $= 0.1714 \times 10^{-8}\,\text{BTUH}/(\text{ft}^2\,{}^\circ\text{R}^4)$

T = Absolute temperature °R (Rankine)

An excellent example of this phenomenon is our planet Earth as the absorber of the radiant energy given off by the sun. No energy is lost in the approximate 93 million mile journey until it comes into contact with the surface of the Earth. It takes less than 500 seconds (8.4 minutes) for the radiant energy to cover that distance. It should be noted that the sun's radiant energy is the primary source of heat for our planet. The molten core of Earth provides very little heat, by comparison, to heat our environment.

This form of heat from the "molten core" is known as geothermal energy. Geothermal energy is the thermal energy in rock and fluid that fills the pores and fractures within the rock sand and gravel. It is believed that the ultimate source of geothermal energy is radioactive decay within the earth.

Through plate motion and volcanism, some of this energy is concentrated at high temperatures near the surface of the earth. Energy is also transferred from the deeper parts of the crust to the earth's surface by conduction and by convection in regions where geological conditions and the presence of water permit.

This information is pointed out to illustrate the cost factors involved to use geothermal energy. The following items influence the cost of energy delivered from geothermal resources: well depth; well flow rate; resource temperatures; and temperature drop.

The cost of the wells is usually one of the major items in the overall cost of a geothermal system, which increases with depth needed for the geothermal project. Well depths in the western United States are relatively shallow. The larger geothermal systems there operate with wells of less than 2,000 feet and many at less than 1,000 feet.

One of the practical applications for geothermal energy is a ground source heat pump, but again you are heating air with the same restrictions. The primary source of a practical means of heating the earth is still infrared wave lengths converted to heat when the infra-red rays meet the earth. Any given body using a fixed set of conditions will generate various wave lengths of radiation. The radiation intensity of the diverse wave lengths is different.

By definition, a "black body" is any object whose surface completely absorbs all thermal radiation that strikes it. Again the thermal radiation emitted by a "black body" is the maximum possible.

Thermal Engineering Corp. developed a Thermal Radiation Calculator, Figure 1.2, which can be used instead of calculating with the above equation. The heavy dashed line shows the total radiation that can be used with their 45,000 BTUH unit (at that time) ceramic surface at 1,730 °F. The overall burner area was 156 square inches with a solid area of 111 square inches with an emissivity of approximately 0.90. Using Chart 1-1, it can be determined that the total thermal radiation is 27,000 BTUH. This calculator can become a valuable sales tool in demonstrating to a potential customer the superior qualities of infra-red heaters. Obtain competitive data on surface area, surface temperature, emissivity, and percent of solid area. Plot them on the calculator and compare with the competition. Fortunately, the IR heater manufacturer must determine these data based on the surface temperature produced by its typical IR heater. In the case of high-temperature IR heaters, the surface temperature of the emitter is somewhere between 1,200 and 1,750 °F. Therefore, the emissivity of these units is approximately 0.90.

As a result of this interchange of radiation and absorption, there is a continuous reciprocation of energy between all bodies.

References

[1] E.R.G. Eckert and Robert M. Drake. Heat and Mass Transfer. McGraw Hill, 1959, p. 358.

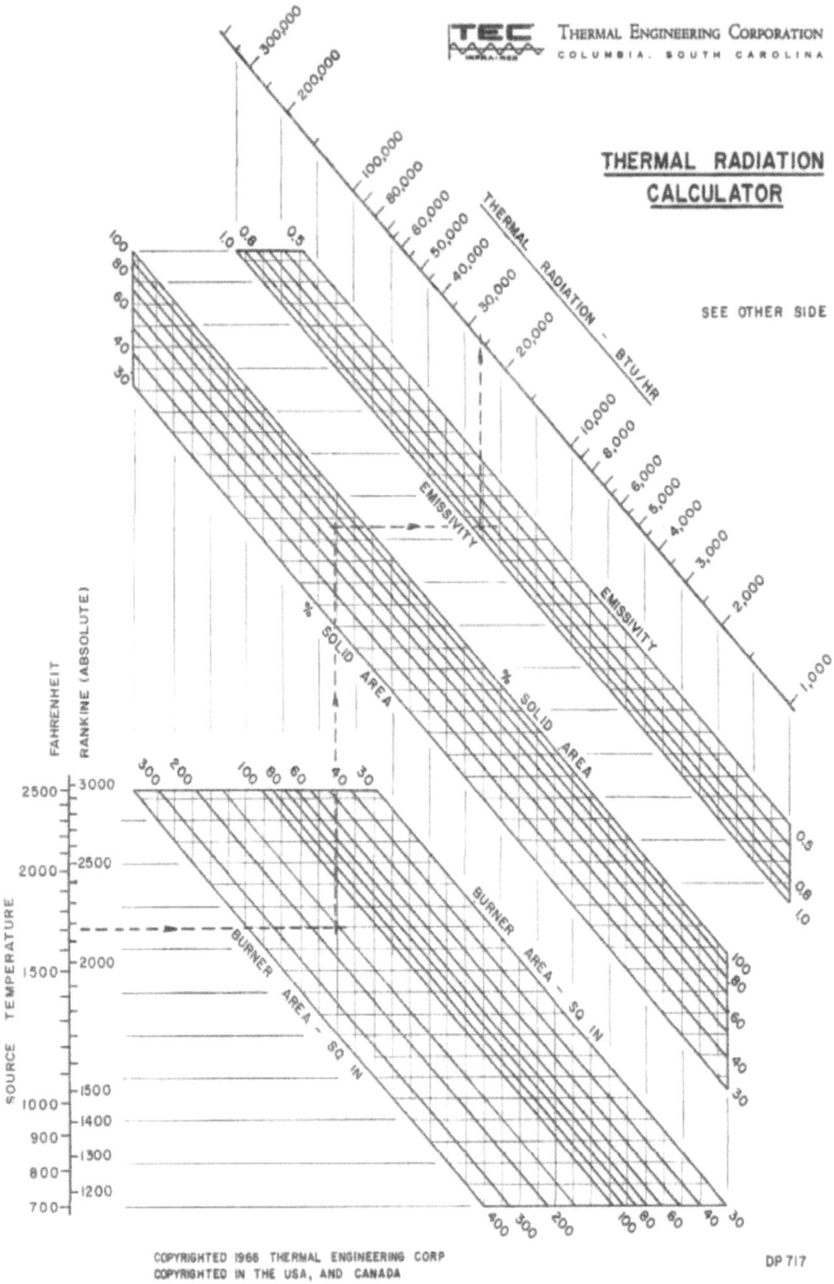

Figure 1.2 Thermal radiation calculator.

2

Evolution of Gas-Fired Infra-Red Heaters

When the war was over, Dr. Schwank emigrated to the United States. He was able to get financial backing to manufacture his originally designed gas-fired infra-red heater at a facility in Cleveland Ohio.

Figure 2.1 shows a typical cross section of a high temperature infra-red burner. Gas is introduced in the proper required amounts through a metering orifice at the gas pressure specified for the heater. Combustion air is aspirated by the gas jet. The gas/air mixture is mixed in the venturi. This mixture enters the burner plenum chamber and is distributed evenly to the inside surface of the ceramic.

The heart of any high-temperature gas-fired infra-red unit is its ceramic. By improving the design of the aperture sizes and the number of apertures per square inch, we have seen major improvements since the first infra-red ceramic devices were introduced. These improvements are:

1. Use of improved apertures for increased flow per unit area.
2. Retrogression of flame below ceramic surfaces for maximum transfer of flame heat into the ceramic.
3. Use of high thermal conductivity coatings on the low conductivity ceramic for rapid surface heat transfer and improved heat erosion resistance.

It follows that the solid ceramic is the primary radiant source; therefore, the number of apertures should be a minimum.

The mixture passes through the circular apertures and burns on the outside surfaces in the divergent passages. A temperature of approximately 1500–1750 °F is developed in a minute or two after the burner is ignited.

Some heater designs require or use a nichrome wire screen positioned about $\frac{3}{4}''$ away from the ceramic surface. It is equal in area to the ceramic area and held in place so that it is free to expand and contract. Its purpose is to speed up heating time, and increase the ceramic temperature. The bright,

DOI: 10.1201/9788770046589-4

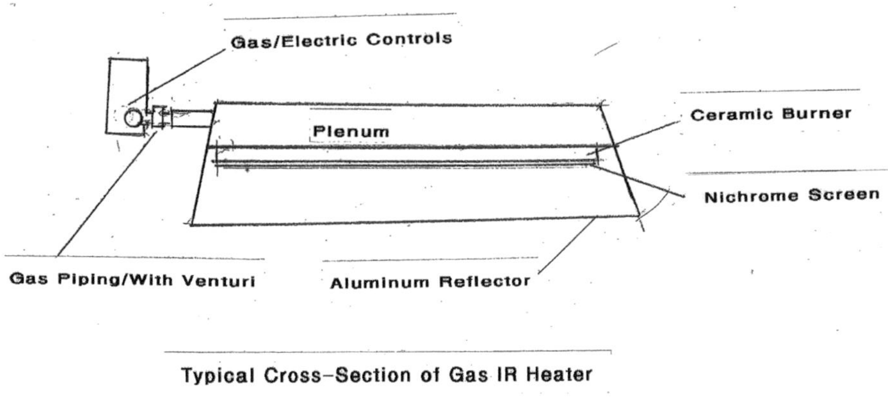

Figure 2.1 Conventional System.

polished aluminum reflector helps direct the infra-red rays toward the target area.

Additional features of infra-red are:

1. No need for interior burner baffles.
2. Ceramic mat is encased in a metal frame and can be replaced in the field.
3. Venturi casing keeps gas/air mixture cool, thereby improving air aspiration characteristics.
4. Burner housings are fabricated of corrosion-resistant aluminized steel, sometimes stainless steel.
5. Circular apertures have self–cleaning ability. Dust, etc., has much larger areas to pass through.
6. Nichrome screens operate below the critical temperature of the heater.

See Figure 2.1.

2.1 Perfection-Schwank

The Perfection-Schwank company was created to manufacture what was the first gas-fired infra-red heater available in this country. For many years, Perfection-Schwank was the only infra-red heater available.

Production was then begun to produce natural gas infra-red unit heaters. The management at the Perfection-Schwank facility was very optimistic with this new form of heating equipment. A sales force was established to help develop manufacturer's representatives to sell these new systems. Efforts were made to train the manufacturer's reps, but specific design procedures were not yet developed. As a result, heating system installations became ineffective as a result of this lack in overall design. The manufacturer's reps were anxious to sell the equipment to make money, which resulted in questionable design recommendations.

As a result, some of the first installations were poorly designed and created a bad reputation for the gas infra-red industry. Fortunately, there were some well-designed installations that kept the business growing. But still there were cases where some heaters with too great a gas input were installed too far apart causing hot spots and some not so hot spots. In some cases, the heaters were mounted too low, causing uncomfortable situations. Another problem with some poorly designed systems was that they sometimes created condensation problems. In other cases, the condensation problems were so severe that condensation literally created "rain" in the heated areas.

It was quite obvious at this time that something had to be done. Manufacturer's representatives wanted to sell heaters, not carefully designed gas-fired infra-red systems. This concept had to be overcome. In some cases, the plant manager or plant engineer tried to solve a "spot" heating problem or used heaters with too great an input mounted too low for the heating problem.

That was when people like Fred Prince, Bill Best, and John Purkey started to create design procedures.

In order to develop this important design data and equipment information, it took much work and extensive field studies. Minimum mounting heights and equipment spacing were determined in laboratory tests and field experiments to develop these design data. Careful measurements were taken to create these design requirements. As a result of this research work and the development of various curves, graphs, nomographs, etc., infra-red heating design is now a practical and workable science.

These created curves, graphs, minimum mounting heights, nomographs, spacing requirements, etc., now corrected these adverse conditions. They used these "tools" to create specification sheets on their various heater sizes. These data included: heat input (Btu/h), recommended mounting height, recommended spacing, and accessories available for the heaters. With the graphs, curves, and nomographs, they were able to train its manufacturer's representatives on how to design effective infra-red heating systems to sell its equipment. For the most part, the training sessions, equipment specs, and literature describing the various units helped overcome the bad reputation made by some of the manufacturer's representatives in the past.

As a result of these design improvements and the physical make-up of these high-temperature infra-red heaters, they qualified as unvented heating equipment. AGA, ASHRAE, and other similar agencies agreed on the unvented status of high-temperature gas-fired IR heaters.

2.2 Roberts-Gordon

It was problems like these that prompted the designers at Roberts-Gordon to come up with the concept of low-temperature vented gas-fired radiant tube infra-red heaters. With the low-temperature tubes, it was relatively easy to provide the means to vent these units to the outside and at the same time eliminate the problem of condensate in the area being heated.

By the late 1950s, Roberts-Gordon pioneered the concept of gas-fired, low-intensity infra-red heaters. This was accomplished by burning an air–gas mixture inside a metal tube or enclosure, which generates infra-red energy and radiates this energy to the space to be heated. The products of combustion are generally exhausted to the outside. Typical operating surface temperatures of the Roberts-Gordon infra-red heaters do not exceed 1,200 °F. The original Roberts-Gordon low-intensity units work on the concept of vacuum-assisted burners to exhaust the products of combustion. In fact, Roberts-Gordon called its low-intensity infra-red heater Co-Ray-Vac due to its vacuum-assisted burner. These vacuum-assisted burners come in two distinct varieties: the condensing and non-condensing types. A vacuum pump system is a complete combustion system consisting of a vacuum pump (at the end), burner(s), control panel, thermostat(s), and 4-inch OD steel tubing for heat exchanger surface in the form of radiant and tailpipe plus assorted reflectors and other hardware. The number of such systems required is based primarily on the heat loss of the building.

These vacuum-assisted burners come in two distinct varieties: the condensing and non-condensing types. Co-Ray-Vac has since introduced a positive pressure tube heater to complement its original vacuum type unit. This unit is also designated as an unvented heater.

As Roberts-Gordon has stated, “the radiant heater which pioneered the transition from warm air heat to ‘heat on the feet’ comfort, provided uniform comfort heat with its low-intensity infra-red heater.” This describes the infra-red concept perfectly in its application.

Air for combustion is supported by room air and can also be ducted from the outside. The unit can be operated in areas of negative air pressure. The Co-Ray-Vac claims that no moisture is released into the building.

Since then, there have been many additional vented and unvented low-temperature radiant tube gas IR heater manufacturers.

2.3 Space-Ray Corp.

One manufacturer came in with a relatively low-temperature high-intensity infra-red heater that was originally used to dry and cure tobacco. That was the Space-Ray Corp., based in Charlotte, NC, which used an atmospheric ring burner (Bunsen-type burner) with an inverted double stainless steel cone. The outer cone has hundreds of tiny perforations to allow the ring burner flame to cause the cone to generate infra-red heater. They chose to have the stainless steel emitter operate in the range of 1000–1200 °F. This unit, designated Model SRO-40, is shown in Figure 2.2. This is considered to be in the low-intensity range. If a higher emitter temperature was needed, the unique way of squeezing the hot combustion gases through the perforated steel emitter was found to be not as efficient because less heat would be squeezed out of the gases before being released. They eventually entered the building heating arena with this Model SRO-40 unit. Space-Ray eventually designed a high-temperature ceramic burner to generate high-intensity infra-red energy. These high-intensity units incorporated stainless steel plenum chambers where the air/gas mixture is produced for combustion. Most of the other manufacturers use aluminized steel for their plenum chambers. The high-intensity units are designated as the RCSA series. Both types of units are still available from Space-Ray Corp. See Figure 2.3. They have also developed low-temperature tube heaters to round out their equipment inventory.

2.4 Thermal Engineering Corp. (TEC)

Another innovator for infra-red heating was Bill Best of Thermal Engineering Corp. (TEC). While teaching and doing research at the University of South Carolina for his doctorate in engineering, where he conducted rocket research for Pratt & Whitney, he fired a flame on a ceramic plate, which was used as a flame arrestor. The temperature of the plate reached over 2000^0F and was glowing and emitting radiant heat. Best could feel the heat radiating from the ceramic plate on his face. This is when he got his first idea for making a radiant burner and infra-red heater. This is when he founded Thermal Engineering Corp. (TEC).

Figure 2.2 Space Ray High Temperature IR Heater (RCSA Series).

Further research with infra-red led him to develop many products using his patented ideas. These ideas included using infra-red for: space heating; paper dryers; automotive paint curing systems; use of infra-red as an aid in the manufacture of radial tires; and infra-red for broiling meats in restaurants and backyard barbecue units. With all these industrial, commercial and now residential uses for his gas-fired infra-red burners, TEC got out of the gas

Figure 2.3 Solaronics Low Temperature IR Heaters.

infra-red building heating business in 1981. Now TEC's primary business is custom designed industrial and commercial process heating applications and gas infra-red grills for backyard barbecues, restaurants, and some fast-food shops. Bill Best's process heating will be discussed in a later chapter.

2.5 Solaronics

An interesting outgrowth in gas infra-red was with the merger of the two companies that were vaguely connected with building heating. These two companies created Solaronics. It was a joint enterprise of Southern Industries, Inc., a manufacturer of stoves and heating equipment since 1857, and Antargaz, S.A. of France. Antargaz was a subsidiary of the largest privately owned French petroleum firm with processing and distributing capabilities throughout the world. The result of this was the formation of Solaronics in 1962. At this time, Solaronics' high-intensity infra-red heating was introduced to the USA and Canada by Solaronics. Eventually, Solaronics would branch out with low-intensity tubular gas infra-red heaters. See Figure 2.3.

2.6 Advances in Infra-Red

American ingenuity was able to come up with additional modifications to both high- and low-intensity IR heating equipment. As a result, there are basically three different types of IR heaters.

1. The original ceramic high-intensity IR units.
2. The metal tube low-intensity units come with either forced draft or vacuum induced draft burners.
3. Finally, low-intensity IR catalytic oxidation heaters are for combustible problem areas.

This book will address primarily the high-intensity and low-intensity IR systems. Low-intensity infra-red catalytic oxidation burners have very limited applications.

In order to develop the utilization of infra-red as a heating source, this text will start by discussing the theory of thermal radiation and then elaborate on a basic design. The design will include basic heat loss calculations, equipment selection and sizing, equipment arrangement, pipe sizing design, installation, and control systems.

2.7 General Advantages

The use of radiant heat or infra-red energy is superior to conventional heating systems for a great number of industrial or commercial applications. This is because gas-fired infra-red energy is:

1. Faster than conventional heating systems because it is instantaneous– there is little or no warm-up period.
2. Heats objects, people, and surfaces, not the air.
3. Infra-red is absorbed into objects and the floor, creating a heat reservoir.
4. Infra-red energy is not normally affected by air movement.
5. The operation and construction of infra-red setups are simpler, and therefore, often more economical than conventional heating systems.
6. Proper design of infra-red equipment produces more uniform radiant flux intensity on the products.
7. Infra-red is re-radiated to occupants and other surfaces of the space.
8. Reduced stratification of air within the space.
9. Reduced actual transmission loss due to lower inside temperature than the assumed design condition as substantially lower ceiling and upper sidewall temperatures due to reduced stratification (25–30 °F lower is not unusual).
10. Infra-red energy can be focused, reflected, or polarized according to the law of optics.

In conclusion, economic analyses will be developed to show how IR systems can reduce overall annual heating costs, compete against other heating systems and equipment, and reduce maintenance costs. The plant manager, design engineer, sales engineer, and the environmentalists can all benefit to reduce energy usage, and heating costs and improve the environment. This is a condition where red, especially infra-red, is green.

2.8 How Space Heating is Accomplished

As mentioned earlier, infra-red energy does not heat the air directly–whether in space or in a building. However, the radiant energy converts to heat energy only when it strikes an object. This phenomenon allows radiant energy heaters (generators) to be used very effectively to provide space heating. There are three basic methods in which heat may be transferred for space heating: convection, conduction, and radiation.

With conventional heating systems, air is used as the primary means to transfer the energy from its source to the space being heated. In a "conventional" system, conduction, convection, and to a small degree radiation are used. Energy from the flame is transmitted by conduction, convection, and radiation through a heat exchanger, usually air or water. This heat energy is then introduced to the area to be heated by convection.

When this heated air passes over an individual at temperature less than body temperature, it will cause a cooling sensation to that person. Therefore, when convection is used to provide comfort heating, it is necessary that the convected air leaving the heat exchanger be at a relatively high temperature to prevent a chilling effect on the occupants of the building.

The higher temperatures needed to provide this condition makes the air buoyant or lighter than the surrounding air and causes it to rise to the ceiling. Because of this phenomenon, where convection heat is used, the temperatures at the top of the building (ceiling) are much higher than at floor level. See Figures 2.4 and 2.5.

With gas-fired infra-red heaters, there are, in effect, two heat exchangers used for the distribution of the energy generated. The emitting surface of the generator is a radiant heat exchanger that transmits the infra-red energy to the floor and other objects in the surrounding area below the infra-red heater. This energy is absorbed by these various objects and instantly converted to heat.

The air close to the floor and various objects that have absorbed the infra-red energy is now heated. In effect, the floor and various objects and equipment in the building become a very large heat exchanger, reflector, and re-radiator for the IR energy. See Figures 2.4 and 2.5. The floor, in effect, becomes a "heat sink". Figures 2.6 and 2.7 illustrate more effectively how the heat is distributed with each form of heating.

There is an important difference between IR heating systems and conventional heating in that the floor and objects that become heat exchangers are heated to a much lower temperature than the heat exchangers in a convection heater. Any air circulating close to the floor or any objects that have absorbed the radiant energy cannot be warmer than the temperature of the objects themselves. As a result, the air heated in an infra-red system is not as buoyant as in a convection system. Due to the fact that the air cannot be any warmer than the heat exchanger from which it gains its heat, the heat is concentrated in the work space instead of rising to the roof.

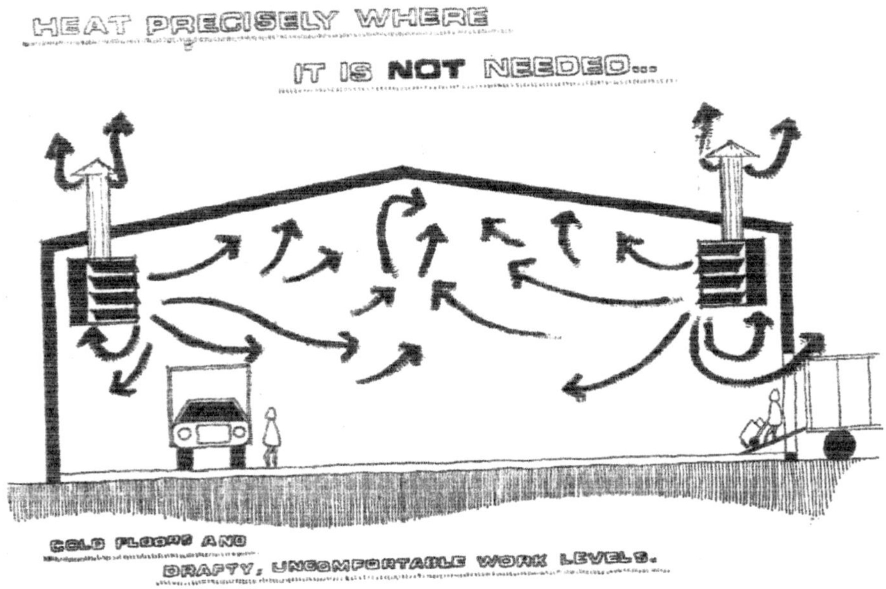

Figure 2.4 Conventional System.

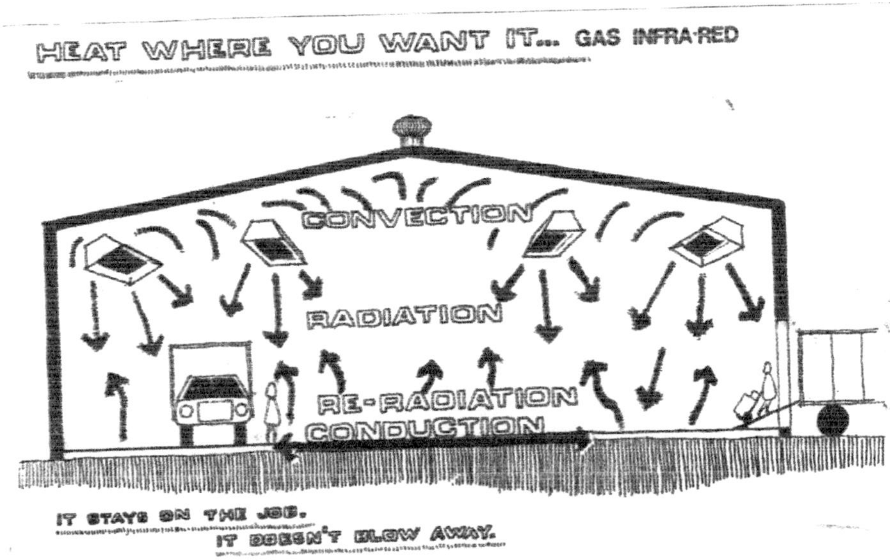

Figure 2.5 Infra-red heating system.

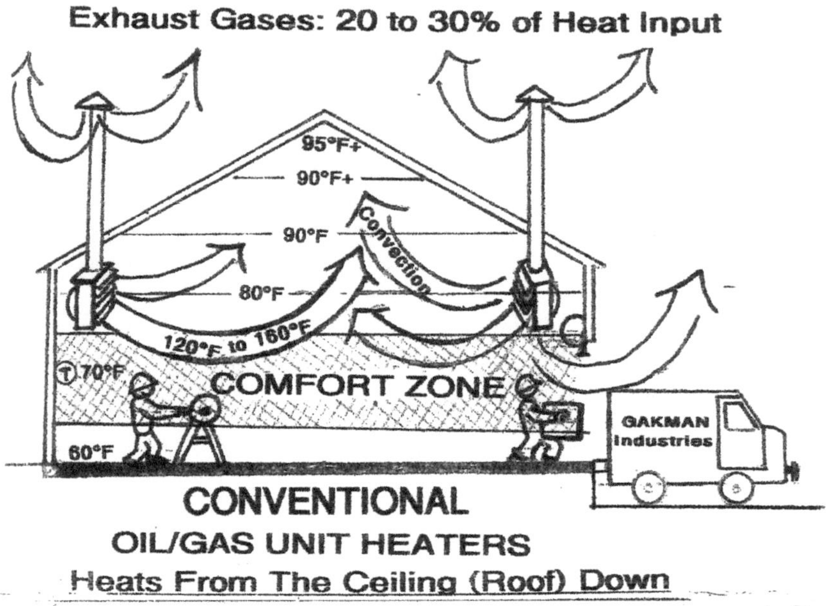

Figure 2.6 Conventional oil/gas unit heaters.

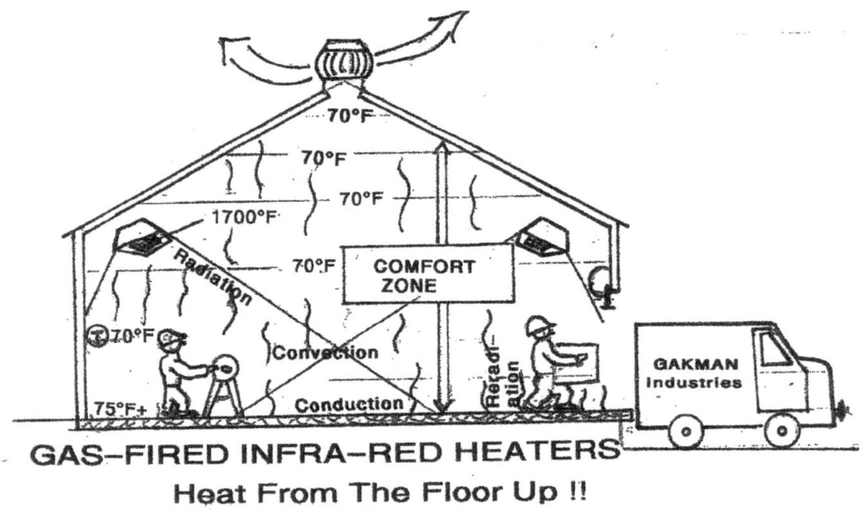

Figure 2.7 Gas-fired infra-red heaters.

2.9 Additional Advantages

The warmest area in buildings heated with an infra-red system is the floor. This is a very important aspect of infra-red effectiveness. When a person's feet are warm, they feel warm. Conversely, when their feet are cold, they definitely feel cold, In a building heated by a convection system, the warmest area is at the ceiling (roof). The floor is the coldest. This is why the heat loss through the walls and ceilings of the infra-red heated building is decreased due to the lower ambient temperatures at the ceiling and upper walls.

Another factor that must be considered is that infra-red energy is provided by radiant distribution. It does not require high-velocity air circulation, which is needed with conventional heating systems that usually introduce an evaporative cooling effect–especially with the entrained air in the primary convection air supply not heated directly by the unit heater.

The byproducts of combustion from the unvented infra-red heaters are buoyant and will rise toward the ceiling and help offset the ceiling or roof heat losses. The temperature at the ceiling will be a function of the mounting height of the infra-red heaters. Variations in the installed heater height as it relates to the ceiling height are illustrated in Curve B shown in Figure 4.3. The use of Curve B will be explained further in the text.

It can readily be seen that lower ceiling temperatures will result when the IR heaters are installed at lower levels relative to actual ceiling heights. IR heaters mounted at lower levels allow the exhaust gases to dilute as they mix with the surrounding air and lower their temperature before they rise to the ceiling. This results in exhaust gases being contained in greater volumes of air, thereby reducing ceiling temperatures.

One-half of the total energy input to the IR heater is for its radiant energy, which is available to offset heat losses of the floor, lower walls and infiltration. In other words, IR energy emitted from the IR heaters does not heat the air first. This results in the conservation of heat. Additional heat is also saved, to a lesser degree, because internal wall and ceiling surface coefficients are decreased due to reduced air movement of IR heaters.

The resultant reduction in air movement lowers the U-value, which, in effect, provides a measure of insulation without changing construction.

Experience has shown that where occupants of an area are exposed to radiation from IR heaters and/or the reduction of radiation losses to warmer floors, walls, and surrounding objects, lower ambient air temperatures can be used, and in most instances insisted upon. These facts further reduce energy consumption and if anticipated can result in lower initial system/equipment costs.

One need not be an Einstein or a rocket scientist to understand that IR systems are more energy efficient and less costly to operate than conventional heating systems. All that is needed is basic common sense and an understanding that heat loss is a function of the difference between outdoor temperature and indoor temperature.

With conventional systems, the ceiling temperatures can be anywhere between 100 °F to more than 130 °F, while it is only 70 °F at thermostat level. It can easily be seen that the heat loss is much greater than was calculated at the 70 °F design level. Some Plant Managers/Engineers install "Casablanca" fans to bring that heat down to the lower levels. All they are really doing is introducing air movement with the resultant evaporative cooling effect. The proof of this is when you ask them what they do in the summer time when it is hot in the building. Their answer is, they turn the fans on to cool the workers. This seems like an oxymoron. You cannot have it both ways. You either heat or cool the people under either condition. It also seems that the roof and/or the ceiling would be much hotter during the summer than it would be at any time during the heating season, especially with solar gain during the "dog" days of summer.

2.9.1 Conventional heater efficiency

Conventional heaters must vent a minimum of 20% of its heat input but more likely 30% in order to keep the water vapor of exhaust gases from condensing and running back into the heater itself causing oxidation and corrosion. The reason for this oxidation and corrosion is that the byproducts of combustion include not only water vapor and CO_2 but NO_x as well. When the NO_x combines with the water vapor, the result is a form of nitric or nitrous acid, which is highly corrosive. With oil-fired units, the corrosion effect is even greater with sulfur in the oil. SO_x combining with any water vapor creates sulfuric or sulfurous acid along with the nitrous acid. This means a maximum efficiency of 80% or realistically an efficiency of 70%. IR heaters are usually unvented, with condensation dealt with in the design. IR heaters are officially rated at 92% overall thermal efficiency.

2.9.2 Electrical considerations

A 200 ft × 80 ft building with a 24 ft ceiling, which will be used as an illustrative problem later in this book, would require a minimum of six unit heaters. Each unit will require a 1/4 or 1/2 horsepower fan for its operation. Assuming 1,500 full-load hours for 1/4 horsepower fans at a typical $0.15 per kW h cost, the annual cost per heater would be:

$$1{,}500\,\text{h} \times \frac{1}{4}\text{hp} \times \frac{\$0.15}{\text{kW h}} \times \frac{746\,\text{W}}{\text{hp}} \times \frac{\text{kW}}{1{,}000\,\text{W}} = \$41.9625$$

$$\$41.9625 \text{ per heater per year} \times 6 \text{ heaters} \approx \$251.78 \text{ per year.}$$

Ten low-intensity, tube-type, IR heaters would be needed. Each unit uses 1/25 HP blower motors for operation. Using the same values as above:

$$1{,}500\,\text{h} \times \frac{1}{25}\text{hp} \times \frac{\$0.15}{\text{kW h}} \times \frac{746\,\text{W}}{\text{hp}} \times \frac{\text{kW}}{1{,}000\,\text{W}} = \$6.714$$

$$\$6.714 \text{ per heater per year} \times 10 \text{ heaters} = \$67.14 \text{ per year.}$$

With the high-intensity heater, the electrical requirements are even lower. They are equipped with a spark ignition requiring 5–6 VA, which is intermittent and drops off after the gas pilot proves to be burning. The annual cost for these 5–6 watts igniters is less than $10.00 per heating season.

2.10 Conventional Heating Systems

To illustrate why gas infra-red is more effective than the other heating systems available, it is necessary to review the competition:

- Gas-fired unit heaters
- Oil-fired unit heaters
- Steam-powered unit heaters
- Gas-fired boiler with radiators
- Oil-fired boiler with radiators
- Gas-fired duct heaters
- Oil-fired duct heaters
- Gas/oil-fired make-up heaters

2.10.1 Gas-fired unit heaters

Manufacturers of gas-fired equipment invariably show gas inputs and equipment outputs with at least 80% thermal efficiency. They do concede that 20% of the gas input must be exhausted to prevent the water vapor of combustion from condensing in the exhaust flue pipe. Should there be condensate in the flue pipe, it would combine with the NO_x and CO_2 to form corrosive acids that would slide back to the unit and eventually destroy the equipment. In the real world, the burners are usually set to have 30% of the exhaust gases go up the flue to make sure no acids will be created as a direct result.

All unit heaters consist of a gas pressure regulator, gas burner, heat exchanger, blower fan, and exhaust flue.

Gas burners for unit heaters come in a wide variety of sizes, shapes, and designs. There are various basic types: slotted port, single port, ribbon, impingement target, and jet. Most of these burners are classified as atmospheric burners. There are in cases of high Btu/h input burners that use power burners where combustion air is under pressure from a separate fan. An 80% efficiency is possible with the power burner.

There are two other unit heater burners that can possibly achieve 80% efficiency. They are forced-draft and induced-draft burners. The forced-draft burners receive its combustion air at pressures above atmosphere pressure from a fan located in the primary air supply. See Figure 2.8. An induced-draft burner has combustion air pulled through the heat exchanger by a vacuum created by an exhaust fan located in the flue. See Figure 2.9.

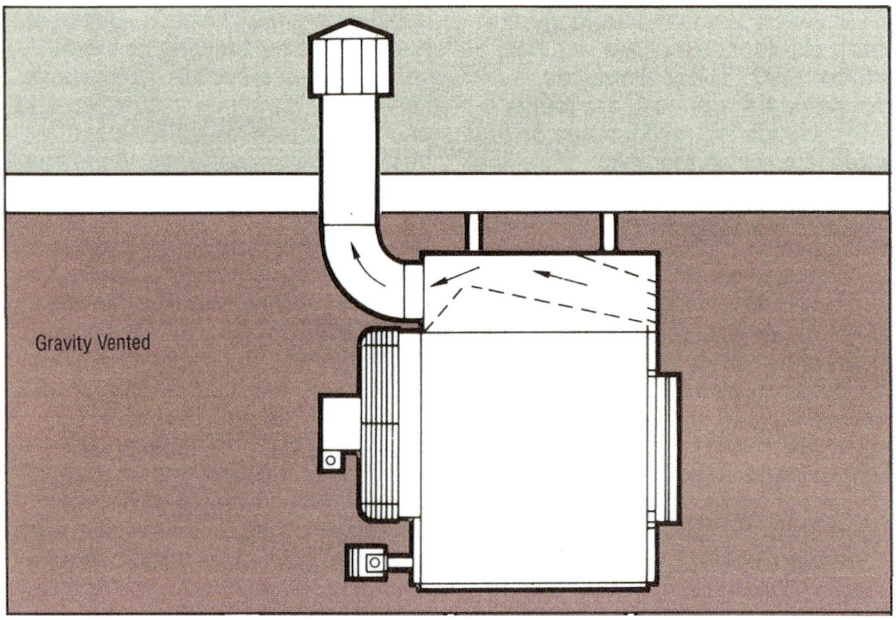

Figure 2.8 Gravity Vented Gas Unit Heater.

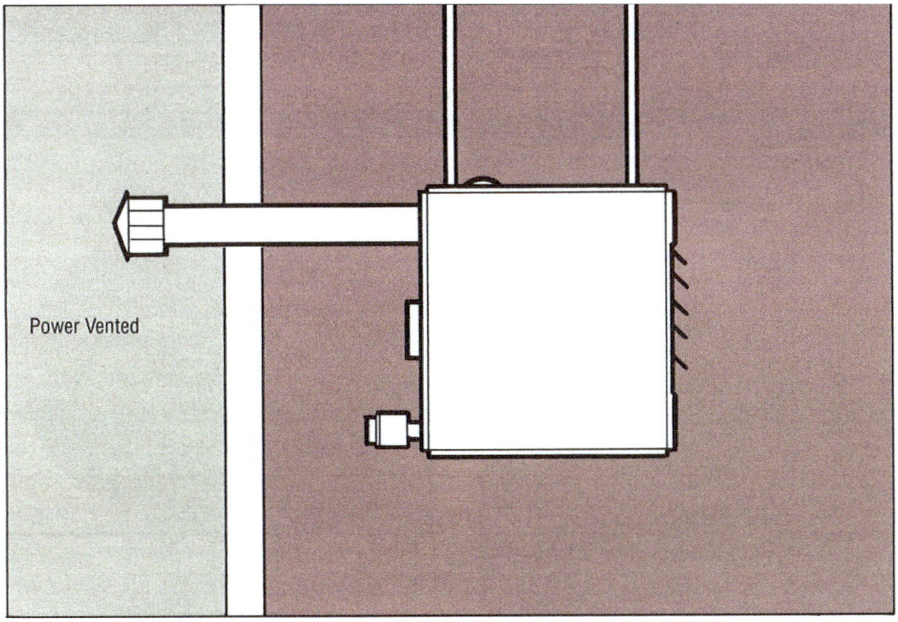

Figure 2.9 Power Vented Gas Unit Heater.

2.10.2 Oil-fired unit heaters

Oil-fired unit heaters have basically the same components as the gas-fired unit heater except for the burner. Instead of an atmospheric burner, it has a fuel pump to force the oil through the burner orifice to atomize the oil for combustion. In order for any fuel to burn, it must be in a gaseous state. Even coal or wood must be heated before they become flammable. An oil-fired unit heater (or boiler) requires a gas pilot light, glow coil, or some other outside means to ignite the atomized oil. It usually takes a few minutes for the burner to heat the heat exchanger efficiently. It is during the heating up period that the burner will be producing soot or lampblack, which will adhere to the exchanger. Obviously, this soot build-up will affect the thermal efficiency.

According to the 5th edition of Marks Mechanical Engineers Handbook, it lists the "Thermal Conductivity of Miscellaneous Solid Substances": the "k" factor for lampblack or carbon at $104\,^{\circ}\mathrm{F}$ is $0.038\,\mathrm{BTUH}/(\mathrm{ft}\,^{\circ}\mathrm{F})$, which comes out to an R-value of $26.3\,\mathrm{BTUH}/(\mathrm{ft}\,^{\circ}\mathrm{F})$. The table does point out: in general, "k" varies with temperature. In the case of metals, the heat exchanger, "k" usually decreases with rising temperature. This means the R-value would be even higher. The reason for this explanation is that the oil industry makes claims that oil is more efficient than gas, with a thermal efficiency of 80%. The unit heaters or boilers may start at the beginning of the heating season at the most of 75%; but you can see how it would drop off quite dramatically.

2.10.3 Steam–powered unit heaters

Steam is no longer an energy specified by HVAC design engineers or architects for building heating. One of the reasons for this is that it is difficult to control for multi-zone heating. Steam systems with condensate returns require steam traps, which always present maintenance problems. When steam traps fail, there is either no heat or too much heat depending on the type of failure. Repair or replacement of steam traps is not only inconvenient but also quite expensive.

Without condensate return, the efficiency of the boiler drops off dramatically. The reason for this is that large quantities of make-up water are required. This make-up water, at or below room temperature, needs quite a bit of heat to bring it into steam at 212 °F. This requirement definitely affects the boiler's thermal efficiency. Providing space for a boiler is another reason for not using steam. There are installations that have steam energized unit heaters and duct heaters. Overall poor efficiency of these systems, along with high maintenance costs present little competition for gas infra-red systems that qualify for radiant heating.

2.10.4 Hot water fueled unit heaters

Oil- or gas-fired boilers provide the hot water required for hot water unit heaters or duct heaters. In the case of any boiler, whether gas, or oil-fired, a minimum of atleast 20%–30% of the fuel input for firing must go up the chimney or flue to prevent the water vapor from condensing and combining with the NO_x and/or CO_2. The result would be nitrous and/or sulfurous acids, which could corrode the boilers.

In nature, no material can burn until it is in a gaseous state. Even steel burns if it gets hot enough. In the case of oil, the burner forces the oil through a carefully sized orifice that atomizes the oil to a fine spray that is ignited by a gas pilot or glow plug. This action causes air inspiration to create an air/oil vapor ratio for complete combustion. On a cold start, there is much soot created until the fire box gets up to temperature. This soot (carbon) will begin to coat the fire tubes. A gradual build-up of the carbon will have a negative effect on the thermal efficiency. By the end of the heating season, the boiler efficiency is about 60%. Coupling this with the line losses and the cost of pumping the water, the cost of operation becomes high, which again is in favor of infra-red heating.

2.10.5 Blow back

There is an additional problem that occasionally happens with oil-fired boilers, which is called *blow back*. This condition can be a very costly occurrence. Blow back is the result of a vast accumulation of soot to the point that it restricts the flow of exhaust gases where, due to the temperature rise in the flue, the soot build-up suddenly ignites in a mild explosion. As a result of this occurring, soot is blown back, or forced back, into the building. Quite often, schools and commercial buildings have been closed or shut down for extended periods of time, sometimes weeks, to clean and often have to repaint because of extensive soot damage caused by blow back.

2.10.6 Relative cost

For new buildings, the installed cost differential is close. The customer may not like having to give up at least 200 ft^2 of floor space for a boiler room. There is also the cost of the boiler plus the installation. The cost of the piping and the installation of the piping required to provide the supply and return hot water for the unit heaters. In addition to this cost is the cost to provide and install insulation for this piping. Add to these costs the cost of the unit heaters and their installation. Pumps to supply the hot water and installation along with the wiring to operate the heaters and pumps also must be included in the overall costs.

Gas infra-red heating may even be less expensive. In any case, the operating cost savings with the IR heating would more than offset any installed cost differential.

2.10.7 Gas/oil-fired boilers with radiators

Gas-fired or oil-fired boilers with radiators are in basically the same conditions as it was for steam-powered unit heater and/or for hot water fueled unit heaters: they both needed costly and inefficient boiler operation. The installation cost may even be higher depending on the cost of radiators and the installation cost of the radiation equipment.

2.10.8 Gas/oil-fired duct heaters

Duct heaters operate in the same manner as unit heaters. A fan forces the air through the heat exchanger into a duct for distribution to louvers throughout the building. It is very important to size the ductwork properly including a proper aspect ratio.

2.10.9 Gas/oil-fired make-up air heaters

It is very costly and inefficient to heat an entire building with make-up air heaters, whether they are oil- or gas-fired.

Oil-fired make-up air heaters are not recommended because the by-products of combustion are part of the make-up air into the building, which could be a definite problem. In addition, an oil burner does not have the capabilities of a turn-down ratio to meet the varied outdoor air temperatures during the heating season.

Gas-fired make-up air heaters do have the ability to meet these varying outdoor air temperatures, with modular burner capabilities. Even with this advantage, it is still an expensive way to heat an entire building. However, with make-up heaters, it is now possible to recover at least 80% of the waste heat of exhaust and the introduction of outside make-up air with an energy recovery ventilator that is now available with a counter-flow ERV. These units can utilize the room temperature air with its patented counter-flow heat exchanger, or in the case of the exhaust from IR tube heaters, to heat the make-up air to within 2 °F of room temperature. This can now increase overall efficiency to about 98% with the use of the patented energy recovery ventilator (ERV). The manufacturer of this patented ERV can be found on the internet.

The ideal use of gas make-up air heaters is actually with gas-fired infra-red units. Infra-red would take care of the convection heat loss and some of the ventilation losses; while the make-up air heaters would be used for make-up air for manufacturing requirements for ventilation when required.

All of the listed "conventional" heating systems require heating the air for comfort heating. These systems all point to a basic equation governed by the basic rules of physics, chemistry, and thermodynamics. We are all familiar with that equation in high school and college, which is:

$$PV = nRT.$$

Since both n and R are constants, we can in effect simply leave them out of the basic equation. So if we raise the temperature in an enclosed volume, and we must assume that the building is not going to expand, then the pressure will increase. This is good because a positive pressure will keep the cold outside air from coming into the building. However, with a positive pressure in the building, there will be an exfiltration of the heated air to the outside. The real problem occurs when anyone opens a door or an overhead door for shipping or deliveries, then the heated air rushes out of the building. This will cause a temperature drop due to outdoor cold air coming in as a result of the sudden loss of building pressure. The cold air coming in, combined with the hot air going out, must now be heated in addition to the normal heat loss. See Figure 2.2. Air temperatures vary from approximately $60\,^{\circ}$F at the floor level to sometimes more than $100\,^{\circ}$F at the roof or ceiling.

Infra-red heating is also subject to the equation $PV = nRT$, but to a lesser degree. The big advantage with an infra-red heated buildings is that when the doors are opened, the positive pressure is much less due to lower average temperatures, resulting in less heated air to the outdoors and as a result less cold air coming into the building. However, when the doors are closed, the comfort level is restored instantly with the infra-red heaters coming on with the drop in temperature, along with the warm floors. The warm floors act as a heat sink to help maintain comfort conditions. So it can be said that infra-red heating is effective even with the condition created by $PV = nRT$.

3

Basic Heating Design

What is a better way to begin a chapter on basic heating design than to quote from the ASHRAE handbook of Fundamentals in Chapter 8 on Thermal Comfort:

> A principal purpose of HVAC is to provide conditions for human thermal comfort, that condition of mind that expresses satisfaction with the thermal environment (ASHRAE Standard 55). This definition leaves open what is meant by "condition of mind" or "satisfaction", but it correctly emphasizes that judgement of comfort is a cognitive process involving many inputs influenced by physical, physiological, psychological and other processes. This summarizes the fundamentals of human thermoregulation and comfort in terms useful to the engineer for operating systems and designing for the comfort and health of building occupants.

This paragraph from the ASHRAE Fundamentals Handbook tells the basis of study for heating design.

3.1 Fundamentals

All heating designs must be concerned with one primary purpose, which is the comfort of the human body. Comfort is an environmental quality easily recognized when it is experienced. However, it is more difficult to define comfort in such a way that it can readily be used to control the thermal environment in an industrial or commercial application. Comfort is associated with a neural thermal sensation during which the human body regulates its internal temperature with a minimum of physiological effort. The physiological behavior of the human body requires a balance between

DOI: 10.1201/9788770046589-5

the rate of internal heat production (metabolism) and the rate of external body heat loss. The human body provides a unique system of temperature control to regulate these losses by radiation, convection, and evaporation. Extremes of bodily heat loss are not only uncomfortable but actually unhealthy. A person in good health maintains an average body temperature of 98.6 °F. There are four major factors that provide comfort for the human body, which are:

1. Temperature
2. Humidity
3. Air motion
4. Air purity

All of these factors are equally important and will be discussed.

3.1.1 Temperature

The human body is, in effect, a "machine" that works best at certain temperatures. Unlike many machines that can withstand a wide range of differences in the surrounding temperatures, the human body can tolerate only a relatively narrow band of temperature differences. Also, the human body is the only "machine" that has an appreciation of comfort. The average person when seated or at rest will give off approximately 400 Btu/h (their metabolism). That same person while doing heavy work or even just exercising will emit heat at a higher rate and the maximum metabolic rate of a person can be ten times as great as when they are seated or at rest. Heat is given off the human body in basically three ways: evaporation, convection, and radiation. In other words, humidity, air motion, air purity, and radiation.

3.1.2 Humidity

A large degree of heat lost by the human body is lost by evaporation from the skin. Extremely low relative humidity can result in health problems as well as affecting the physical properties of many substances. Under normal comfort conditions, approximately 25% of the body loss will be dissipated in water vapor from the lungs and by sweating from the skin. This 25% or 100 Btu/h for a person at rest will remain relatively constant for comfort conditions. An example of how low relative humidity can be by raising the temperature can be seen using a psychrometric chart. Outside air at 30 °F and 100% RH heated to 70 °F would result in a 21% RH. This air would, in effect, be drier than the Sahara. See psychrometric Chart 3.1.

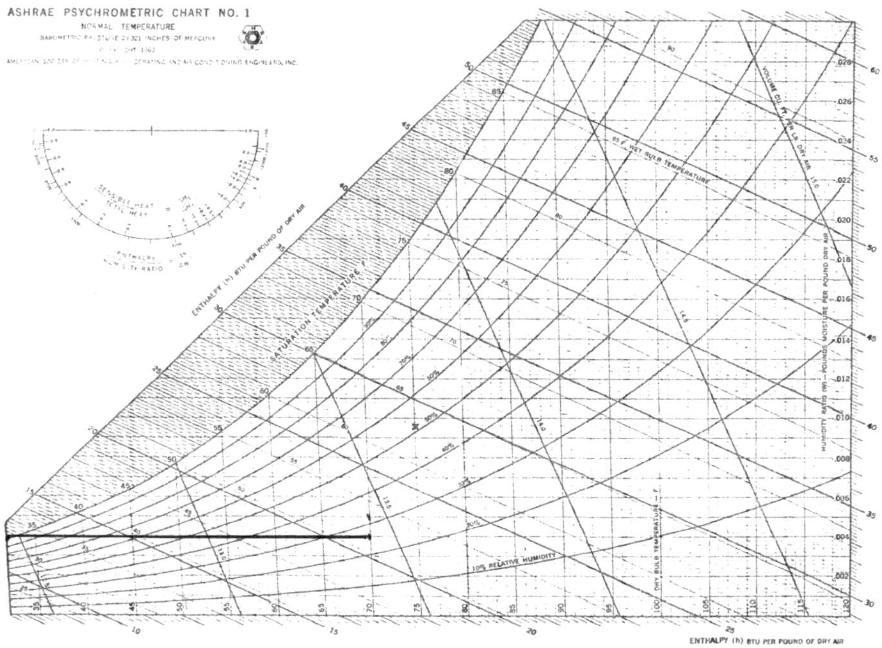

Chart 3.1 Temperature Effect on Relative Humidity.

3.1.3 Air motion

The velocity of air in a building will affect the rate of heat and moisture dissipation from the human body, which is commonly referred to as "evaporative cooling." The body heat loss transferred to the air by convection will be proportional to the temperature difference between the body and the surrounding air. In a room maintained at 70 °F, the convection loss will be approximately 100 Btu/h with still air conditions for a person at rest. This is somewhat acceptable in summer cooling design; however, in winter, low RH can create very uncomfortable conditions with exposure to air motion combined with low RH during the heating season, workers may experience dryness of eyes, nose, and throat, leading to more serious health problems.

3.1.4 Purity of air

General ventilation supplies air to dilute contaminants and replace exhaust air in the workplace. Ventilation can be provided by natural or mechanical supply and/or exhaust systems. The removal of solid particulate matter and odors from air in the controlled space or introduced into the space is very important, especially as it relates to the potential health risk to the human body. Some industrial/commercial facilities require makeup air for safe effective operation to replace the large volumes of air being exhausted to provide comfort and safety for its personnel and/or its process operation. Any ventilation used in a building must be included in the heat loss calculations. In some cases, the ventilation requirement is so great that infra-red may not be adequate to realistically make up the heat loss requirements and may require make-up heaters in your design. This is a condition that was mentioned earlier in Chapter 2 where an energy recovery ventilator could help solve the make-up air dilemma. This counter-flow ERV unit uses the room temperature air to heat the outdoor make-up air to within 2 °F of room temperature, without any make-up air heaters. See Insert 1 on page 7.

3.2 Comfort Parameters Per Metabolic Rate

The "thermal balance" that defines the feeling of comfort is dependent on the way the human body loses its heat to the environment. Everything above absolute zero radiates in all directions. If surrounding objects are at a lower temperature, the radiation will result in a net loss. If, diametrically, the surrounding objects are at a higher temperature, the results will be a net gain. The clothed person with an effective surface temperature of 84 °F will: lose heat by radiation to walls, floors, ceiling, or machinery if they are below 84 °F; or gain heat if the objects are at a higher temperature. In a 70 °F room, a person at rest will lose approximately 200 Btu/h by radiation. Food eaten by a person is turned into energy by an oxidation process with the air breathed into the lungs. The human body as a "machine" has a very low efficiency, which causes it to reject a portion of its energy intake into the environment. The rate in which the body produces heat is known as the metabolic rate. The amount of heat produced varies widely depending on the activity of the person.

3.3 Controlling Body Heat Loss

A person must lose heat to be comfortable. This heat loss will be by evaporation, convection, and radiation. Because evaporation loss must remain relatively constant to maintain comfort, it is necessary to regulate the convection and/or radiation dissipation. As long as the total body heat loss can be kept at its proper level, comfort will result.

Body heat loss by convection can be maintained and varied by adjusting the temperature controls. In well-built, low-ceiling buildings, it is relatively easy to control. However, as the roof/ceiling height increases or as cold air infiltration increases, convection heating becomes difficult to control. As a result, larger and larger volumes of air must be heated to keep acceptable temperatures for the workers.

Spot or partial area heating by convection is not realistic or practical to use unit heater or conventional heating equipment. It is literally impossible to heat air in one part of the building without that heated air migrating into other sections of the building, especially the overhead, since hot air rises.

Lowering the air temperature will result in an increase in convection losses from the workers. This increase in heat loss by convection can be offset by decreasing the radiation loss so that the total heat loss remains the same. This can easily be done by simply installing gas-fired infra-red within the area. The person will continue to radiate to the surroundings but will realize a net increase of radiation from the overhead infra-red heaters.

In low-ceiling and well-insulated buildings, the amount of heat necessary to heat by radiation may be the same as for convection heating. However, as the height of the building increases or as the quality of construction decreases, the need for gas-fired infra-red heating becomes more obvious.

Infra-red heating is not new. Actually, it was the heating concept first known to man. The cave man's only source of heat was the sun. Until man learned how to control fire, it was radiant heat from the flames and hot coals that warmed his fur-clad body. Scientists have long considered the basic properties of infra-red as being ideal for heating but only recently have these principles been put to practical use. Consider the fire place and the Franklin stove.

3.4 Concept of Effective Temperature

It is virtually impossible to design a heating system that satisfies all people. That is because they react in widely different ways to atmospheric conditions around them. What may be comfortable to one person may be "chilly" to another and "too warm" for someone else. Consequently, there is no exact physiological parameter by which comfort can be gauged. As an example, "dry" air in the summer helps dissipate body heat by evaporation (which feels cool); whereas during the winter season, "dry" air can produce an uncomfortable condition due to lowering the skin temperature by evaporative cooling. The degree of warmth felt by the human body depends on the value of metabolic rate, which is always positive since net heat flow is always from the human body to air. Therefore, in order to evaluate the overall effect of evaporation, radiation, and convection, the term "effective temperature" has been designated.

In order to satisfy the comfort requirements of the human body, air temperature, relative humidity (RH) and air motion must act in concert to produce the sensation of warmth or cold. The *effective temperature* is defined as an ordinary index of the degree of warmth or cold felt by the human body in response to temperature, humidity, and air movement.

ASHRAE has conducted a research project over many years, involving thousands of people in different localities in an attempt to relate the factors of temperature, humidity, and air movement to develop its concept of effective temperature. It is determined that for any given air velocity, there were a number of combinations of dry bulb (DB) temperature and relative humidity (RH), which would give the same feeling of comfort to over 90% of the people tested. All people involved in this project were equally clothed and equally active.

Using these, data of DB temperature and RH points were plotted and lines drawn through these points on a psychrometric chart, where the majority of the people tested had the same feeling of comfort, to develop a "Comfort Chart." The Comfort Chart was based on the following assumptions:

1. Subjects clothed for normal indoor living.
2. Subjects limited to no more than light activity, office work, and occasional walking around.
3. Air velocity of 15–30 feet per minute.
4. Radiation effects are neglected.
5. Subject should have normal health.

The results of the tests were plotted for both summer and winter where people feel comfortable. Both zones are fixed within RH between 30% and 70%. The summer comfort temperature zone is between 64 and 79 °F DB; whereas, the winter comfort temperature zone is between 65 and 70 °F DB. See the winter comfort zone shown on the psychrometric Chart 3.2.

3.4.1 Limitations of the comfort chart

There are definite limitations to the comfort chart that must be addressed. The following are considerations not included in the comfort chart determination:

1. The presence of appreciable drafts was not taken into consideration, especially in the heating mode where drafts will cause chart values to lose their meaning.

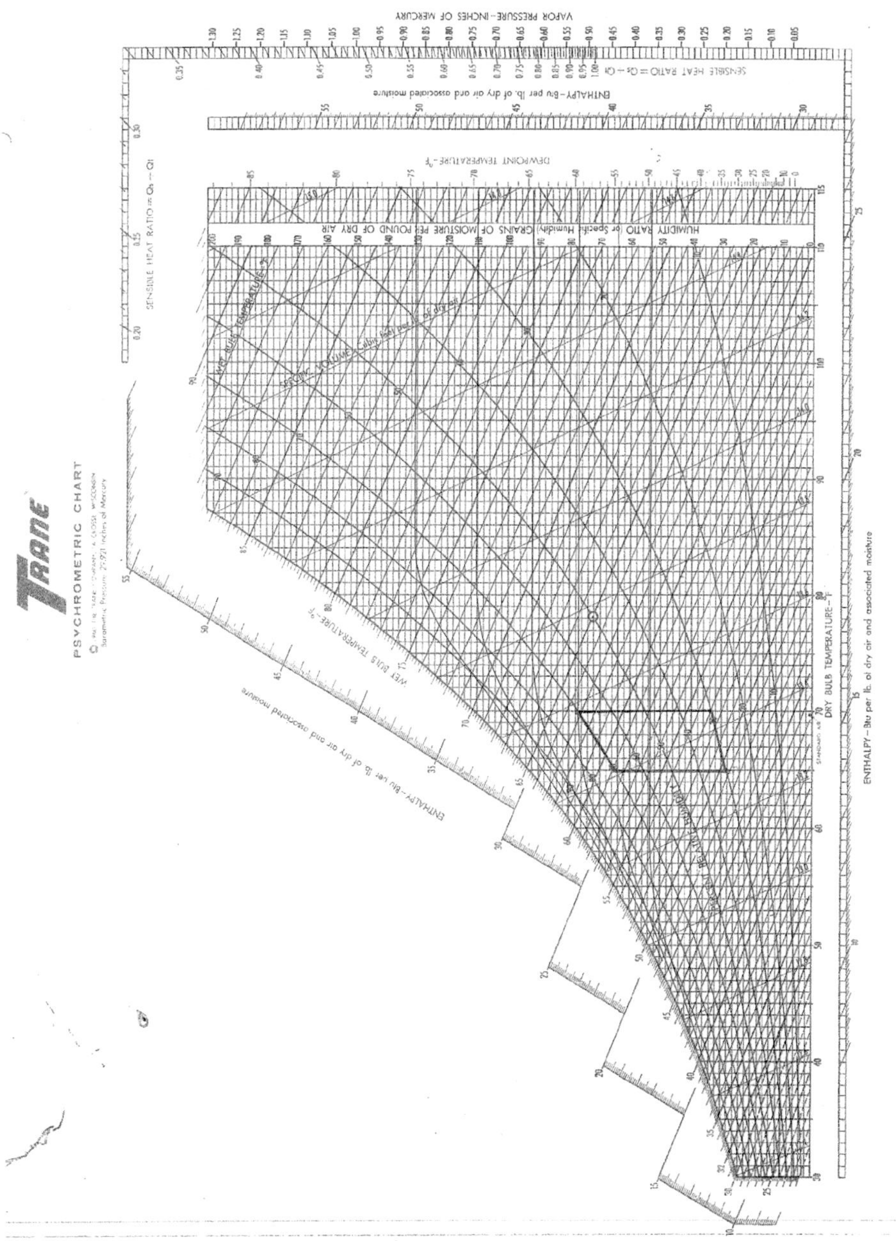

Chart 3.2 Psychrometric Chart Indicating Comfort Zone.

2. Radiation effects were neglected in the preparation of the chart and these factors are often of great significance. During the winter, the human body may radiate so much heat to a cold wall or surface that the skin temperature may cool well below the comfort level even while the room temperature is within the comfort zone.

Coincidentally, there are conditions that have parameters that influence the comfort chart effective temperatures. The more obvious factors that govern the comfort chart effective temperatures are as follows:

1. **Climate differences** People living in North Dakota or other cold climate areas are more comfortable at lower effective temperatures than say people in Virginia/North Carolina climate conditions. The same conclusions occur with RH in Arizona/New Mexico versus Louisiana/Mississippi.
2. **Age and sex** Women require about 1 °F higher temperatures than men. Men and women above the age of 40 usually require an average effective temperature 1 °F higher than those younger than 40. It is tough getting old.
3. **Activity** The degree of bodily activity will have an impact on the effective temperature of the space. Operations with people involved with heavy manual labor require a lower effective temperature than people doing light assembly work.
4. **Clothing** During the winter season, much of the effect due to climate conditions is compensated for because people can be comfortable by just adding and/or removing clothing. Allowance should therefore be made in heating design based on the amount of clothing worn.

3.4.2 Winter design conditions

Any significant difference in the building design temperature from the optimum effective temperature may result in loss of comfort conditions, along with loss of physical as well as mental efficiency. In effect, this means no well-designed heating system installation can justify departure from an acceptable effective temperature.

Any deviation from the optimum effective temperature in the heating system design is not in good judgment except for possible savings in the initial installation cost or the high operating cost during an energy crisis.

4

Heat Loss Calculations

Most or at least some of you have calculated heat loss of a building. However, when calculating heat loss of a building to be heated by gas-fired IR units, the rules have changed.

The rules have changed because gas-fired infra-red requires considerations that need to be addressed that conventional heating systems can overlook. These conditions have to do with fuel input required to maintain design conditions; stratification of heat; condensation possibilities; number of heaters to be effective; correction factors in design; mounting heights; clearances to combustion; electrical requirements; initial installation costs; spot heating; and comfort conditions.

The term *heat loss* has become a common way of describing the heat that is transmitted from the inside of a building to the outside when the outside temperature is lower than the temperature set for inside of the building. A typical building, see Figure 4.1, will be used as an example for heat loss calculations.

This heat flow from the warmth of the building to the colder outdoor temperature occurs through three general areas:

1. Transmission losses - through walls, roofs, and partitions.
2. Transmission losses - through floor slab and edges.
3. Convection losses - cold entering air is heated and carried away through exhaust system's exfiltration.

Transmission losses through walls, roof, windows, etc., are determined by using the following formula:

$$\text{THL} = A \times U \times \text{TD},$$

where:

- THL is the transmission heat loss in BTUH
- A is the area of surface involved in ft^2

DOI: 10.1201/9788770046589-6

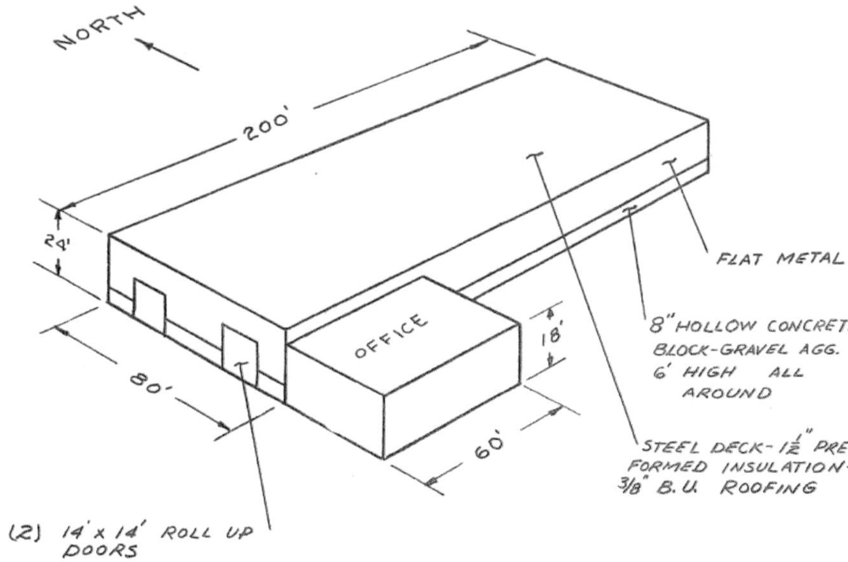

Figure 4.1 Typical Manufacturing Facility.

- U is the overall coefficient of heat transfer in BTUH/(ft^2 °F)
- TD is the temperature difference between desired inside air temperature and outside design temperature in °F.

In order to illustrate an example of heat loss calculations, a typical building will be used. The net area of a surface is obtained from scaled drawings or from direct measurement. The U-values are obtained from Table 4.1 or from the ASHRAE Guidebook. Note that the U-values given in Figure 4.2 are for 15 MPH wind velocity. Conversions of U-values for wind velocities other than 15 MPH are given in Table 4.2.

To calculate U-factors is a relatively simple computation. All materials have an R-value. Many are listed in the Thermal Resistances in Table 4.3. These tables do not list some of the newer materials available in today's markets. However, they all have R-values that are listed with their specifications. The R-values as listed in the tables are shown: as either $1/k$ the value per inch of thickness; or $1/c$ for the thickness listed in the thickness column. In the event the insulating material is new, the R-value can be obtained from the supplier of the material or from the manufacturer's literature or specifications. Once all the R-values are obtained, it is simply a matter of adding all the R-values to get a total R-value. Now it's just a matter of calculating: $U = 1/R_{\text{total}}$.

Table 4.1 Design heat transmission coefficients

THERMAL ENGINEERING CORP.
COLUMBIA, SOUTH CAROLINA

DESIGN HEAT TRANSMISSION COEFFICIENTS
"U" Values (For 15 mph Wind Velocity)

WALLS

INSULATION THICKNESS - INCHES
Mineral Wool, Processed from Rock, Slag, or Glass
Density 1.5 - 4.0 #/ft^3

MATERIAL	0	1/2	1	1½	2	3
Asbestos Cement Bd 3/8" (Transite - Flat)	1.06	0.36	0.22	0.15	0.12	0.083
Asbestos Cement Bd 3/8" (Transite - Corrugated)	1.2	0.37	0.22	0.15	0.12	0.084
Metal - Flat	1.2	0.37	0.22	0.15	0.12	0.084
Metal - Corrugated	1.4	0.39	0.23	0.16	0.12	0.085
Metal - Asbestos Coated	0.9	0.34	0.21	0.15	0.12	0.082
Wood - 1"	0.55	0.27	0.18	0.14	0.11	0.078
- 2"	0.35	0.21	0.15	0.12	0.10	0.072

MATERIAL	THICKNESS - INCHES		
	6	8	12
Brick - Face & Common	0.68	0.48	0.35
Brick - Common Only		0.41	0.31
Concrete Block (solid)			
Gravel Aggregate		0.52	0.47
Cinder Aggregate		0.39	0.36
Lt. Wt.Aggregate		0.35	0.32
Concrete Block (Hollow)			
Gravel Aggregate		0.51	0.47
Cinder Aggregate		0.39	0.37
Lt. Wt. Aggregate		0.36	0.33
Concrete Block - Hollow - 4" with 4" Face Brick			
Gravel Aggregate		0.34	
Cinder Aggregate		0.30	
Lt. Wt. Aggregate		0.27	
Concrete Block - Hollow - 4" with 4" Common Brick			
Gravel Aggregate		0.30	
Cinder Aggregate		0.27	
Lt. Wt. Aggregate		0.25	

DP717

(*Continued*)

Table 4.1 *Continued*

TEC® INFRA-RED THERMAL ENGINEERING CORP.
COLUMBIA, SOUTH CAROLINA

DESIGN HEAT TRANSMISSION COEFFICIENTS
"U" Values (For 15 mph Wind Velocity)

WINDOWS

MATERIAL	"U" VERTICAL	"U" HORIZONTAL
Glass - Single Pane	1.13	1.22
Glass - Double Pane		
3/16" Air Space	0.69	0.75
1/4" " "	0.65	0.70
1/2" " "	0.58	0.66
1" - 4" "	0.56	
Plastic Bubble		
Single	1.09	1.15
Double		0.70

DOORS

MATERIAL	"U"
Flat Metal	1.2
Corrugated Metal (Roll Up)	1.4
Wood 1"	0.64
1¼"	0.55
1½"	0.49
2"	0.43
¼" Plywood	0.86
½" Plywood	0.67

PARTITIONS

MATERIAL	"U"
Flat Metal	0.74
Glass	0.73
Plywood	
1/4"	0.60
3/8"	0.55
1/2"	0.50
Gypsum Board - 3/8"	0.60
Hollow Concrete Blk	
4"	0.45
8"	0.40

FLOORS - SLAB ON GRADE

Uninsulated Edge - 0.81
Insulated Edge - 0.55

AIR

Volume (CFH) x 0.018 x TD

DP717

Table 4.1 *Continued*

DESIGN HEAT TRANSMISSION COEFFICIENTS
"U" Values (For 15 mph Wind Velocity)

MATERIAL	THICKNESS - INCHES			
	4	6	8	12
Concrete - Poured				
140#/ft^3		0.75	0.67	0.55
80#/ft^3		0.31	0.25	0.18
30#/ft^3	0.19	0.13	0.10	

ROOF

MATERIAL	"U"
Metal - Flat - No Insulation	1.3
Metal - Corrugated - No Insulation	1.5
Metal - Asbestos Coated	0.9

MATERIAL	INSULATION THICKNESS - INCHES PREFORMED INSULATION ABOVE DECK 3/8" Built Up Roofing					
	0	1/2	1	1½	2	3
Metal - Flat	0.90	0.40	0.26	0.19	0.16	0.11
Wood - 1"	0.48	0.29	0.21	0.16	0.14	0.10
- 2"	0.32	0.22	0.17	0.14	0.12	0.09
- 3"	0.23	0.17	0.14	0.12	0.10	0.08
Gypsum Slab						
2"	0.43	0.27	0.20	0.13	0.11	.083
3"	0.35	0.24	0.18	0.12	0.10	.079
4"	0.29	0.21	0.16	0.11	0.096	.076
Metal - Flat (Insulation Under Deck-) (Batt Type, Mineral Wool) (Density 1.5 - 4.0#/ft^3)			0.22	0.16	0.12	0.084

DP717

	Material	Size	Area x "U" x TD	BTUH	SUB TOTAL
North Wall	8" Con.Blk	6'x 200'	1200 x 0.51 x 65	39,800	
	Flt.Mtl.	18'x 200'	3600 x 1.2 x 65	280,800	
	Slab		200 x 0.81 x 75	12,200	
					332,800
East Wall	8" Con.Blk	6'x 80'	480 x 0.51 x 65	15,900	
	Flt.Mtl.	18'x 80'	1440 x 1.2 x 65	112,300	
	Slab		80 x 0.81 x 75	4,900	
					133,100
South Wall	8" Con.Blk	6'x140'	840 x 0.51 x 65	27,800	
	Flt.Mtl.	18'x140' 6'x 60'+	2880 x 1.2 x 65	224,600	
	Slab		140 x 0.81 x 75	8,500	
					260,900
West Wall	8" Con.Blk	6'x 52'	312 x 0.51 x 65	10,300	
	Doors Cor Stl	2(14'x14')	392 x 1.4 x 65	35,700	
	Flt.Mtl.	24' x 80-312-392	1216 x 1.2 x 65	94,800	
	Slab		80 x 0.81 x 75	4,900	
					145,700
Roof	Mtl.Deck 1½" Insul B.U.	80 x 200'	16,000 x 0.19 x 65	197,600	
					197,600

TOTAL TRANSMISSION LOSS 1,070,100 BTUH

.85 x .966 = .82

Air Required for Dilution

Total Transmission Loss x Air Requirement Factor

1,070,100 x .225

240,800

ACTUAL AIR EXHAUST

TYPE	CFH
Gravity	
Infiltration	
Process	
Dilution	
Other	

*Total CFH x .018 x TD	Total Air Loss BTUH
240,800 x .018 x 65	281,700

**If the total air exhaust is more than 1-1/2 times the transmission loss, air heaters may be required or the problem can be treated as area heating.*

Figure 4.2 Heat loss calculation.

NET HEAT LOSS

Transmission Loss	1,070,100
Air Loss	281,700
Other	-
Total Loss	1,351,800
Less Heat Gain	-
Net Heat Loss	1,351,800

INFRA-RED CAPACITY REQUIRED

Net Heat Loss x Correction Factor = Infra-Red Capacity Required

1,351,800 x .82 = 1,108,500

Divide IR Capacity Required by Heater Capacity to Determine Number of Heaters Required

1,108,500 ÷ ~~60,000~~ 50,000 = ~~18.5~~ 22

$$\text{FUEL COST ESTIMATE} = \frac{\text{Net HL x Correction Factor x 24 x DD}}{\text{TD x Efficiency}} \times \frac{\text{Cost Per Unit of Fuel}}{\text{BTU Per Unit of Fuel}}$$

$$\frac{1{,}108{,}500 \times 0.82 \times 24 \times 4800}{65 \times 0.92} \times \frac{0.70}{1{,}000{,}000}$$

$ ~~1366~~ 1750.00/yr

$$\text{INLET AIR AREA REQUIRED} = \frac{\text{CFM Exhaust}}{\text{Allowable Inlet Velocity - FPM}}$$

$$= \frac{240{,}800/60}{600}$$

$$= 6.7 \text{ ft}^2$$

Figure 4.2 *Continued.*

Table 4.2 "U" Conversions for velocities other than 15 MPH

"U"	"U" for Wind Velocities - MPH					
15 MPH	0	5	10	20	25	30
0.05	0.049	0.05	0.05	0.05	0.05	0.05
0.06	0.059	0.059	0.06	0.06	0.06	0.06
0.07	0.068	0.069	0.07	0.07	0.07	0.07
0.08	0.078	0.079	0.08	0.08	0.08	0.08
0.09	0.087	0.089	0.09	0.09	0.091	0.091
0.10	0.096	0.099	0.10	0.10	0.101	0.101
0.11	0.105	0.108	0.109	0.11	0.111	0.111
0.13	0.123	0.127	0.129	0.131	0.131	0.131
0.15	0.141	0.147	0.149	0.151	0.151	0.152
0.17	0.158	0.166	0.169	0.171	0.172	0.172
0.19	0.175	0.184	0.188	0.191	0.192	0.193
0.21	0.192	0.203	0.208	0.212	0.213	0.213
0.23	0.209	0.222	0.227	0.232	0.233	0.234
0.25	0.226	0.241	0.247	0.252	0.253	0.254
0.27	0.241	0.259	0.266	0.273	0.274	0.275
0.29	0.257	0.278	0.286	0.293	0.295	0.296
0.31	0.273	0.296	0.305	0.313	0.315	0.317
0.33	0.288	0.314	0.324	0.333	0.336	0.338
0.35	0.303	0.332	0.344	0.354	0.357	0.359
0.37	0.318	0.350	0.363	0.375	0.378	0.380
0.39	0.333	0.368	0.382	0.395	0.399	0.401
0.41	0.347	0.385	0.402	0.416	0.420	0.422
0.43	0.362	0.403	0.421	0.436	0.441	0.444
0.45	0.376	0.420	0.439	0.457	0.462	0.465
0.50	0.410	0.464	0.487	0.509	0.514	0.518
0.60	0.474	0.548	0.581	0.612	0.620	0.626
0.70	0.535	0.631	0.675	0.716	0.728	0.736
0.80	0.592	0.711	0.766	0.821	0.836	0.847
0.90	0.645	0.789	0.858	0.927	0.946	0.960
1.00	0.695	0.865	0.949	1.034	1.058	1.075
1.1	0.742	0.939	1.039	1.142	1.170	1.192
1.2	0.786	1.010	1.129	1.250	1.285	1.318
1.3	0.828	1.080	1.217	1.359	1.400	1.430

As an example, if we had a 1" thick hardwood roof with 1" preformed insulation, the overall U-factor for this roof would calculate as follows using Table 4.3:

Outside air film	0.17
1" Preformed insulation	2.78
1" Hardwood roof	0.90
Inside air film	0.68
Total R value	4.53

$$U = \frac{1}{R_{\text{total}}} = \frac{1}{4.53} = 0.2207 \text{ BTUH/ft}^2/\,^\circ\text{F}.$$

The heat loss for this particular roof would be:

Heat Loss = 0.2207 $\times$ sq. ft. ΔT.

The U-factor for the walls would be calculated the same way for the net sq. ft. and ΔT. U-factors for doors and windows can usually be found in tables or the manufacturer's literature of specifications.

The temperature difference ΔT is obtained by subtracting the outside design temperature from the desired inside air temperature. The outside design temperature can be found in the Appendix – Winter Climatic Conditions.

The outside design temperatures given in Column 5 of the Degree Day Tables listed in the Appendix are reprinted from the ASHRAE Guide and Data Book. Note that three values are given. The following excerpt from the ASHRAE Guide and Data Book will help explain the use of the tables.

Three temperature levels are shown for each location in Column 5. The Median of Extremes is the middle value of the coldest temperature recorded each year for periods up to 25 or 30 years, and is approximately equal numerically to the Average Annual Minimum published in some previous editions.

In the Canadian Tables, the Average Annual Minimum is the value listed. The 99% or 97.5% values represent the temperatures which equaled or exceeded these proportions of the total hours (2160) in December, January, and February. In a normal winter, there would be approximately 22 hours at or below the 99% value, and approximately 54 hours at or below the 97.5 design value.

The Wind data shown in Column 6 pertain specifically to velocities that occur coincident with periods of extreme cold. The four classes of velocity are VL, L, M, and H. These values are Very Light, Light, Medium and High.

Table 4.3 Resistance Values "R"

Asbestos - Cement Board	0.25	Per In. of Thickness
Plywood	1.25	" " " "
Wood Fiber - Hardboard Type	0.72	" " " "
Insulation		
Mineral Wool, Fibrous Form Processed From Rock, Slag, or Glass 1.5-4.0#/ft^3	3.70	Per In of Thickness
Preformed for Use Above Deck	2.78	" " " "
Concretes		
Lightweight Aggregates		
100 #/ft^3	0.28	" " " "
80 "	0.40	" " " "
60 "	0.59	" " " "
40 "	0.86	" " " "
Sand & Gravel Aggregate	0.08	" " " "
Brick Common	0.20	" " " "
Face	0.11	" " " "
Metals - Thin	0.00	
Roofing		
Asphalt Roll Roofing	0.15	For Thickness Listed
Built-up Roofing 3/8 in	0.33	" " "
Slate 1/2"	0.05	" " "
Wood		
Hardwoods	0.91	Per In of Thickness
Softwoods	1.25	" " " "
Film Coefficients		
Inside - Horizontal - Winter	0.61	
" - Vertical - Winter	0.68	
Outside 7 mph	0.25	
15 mph	0.17	

EXAMPLE

$$"U" = \frac{1}{R_{total}}$$

For 1½" Sandwich Construction - Wall

Inside Film	0.68
Metal	0.00
Insulation - Batt	
3.70 x 1½ =	5.55
Metal	0.00
Outside Film	0.17
R_{total} =	6.40

$$U = \frac{1}{6.4} = 0.156$$

DP717

High Velocity is the 15 mph value. Medium is rated as 10 mph, Light Velocity is 5 mph, while Very Light is less than 5 mph. Wind velocities greater than 15 mph will have a definite effect on U-values. Table 4.2 provides correction factors dependent on the value of the U-factor and the wind velocity over 15 mph.

Before selecting a design temperature from the table, the heating engineer should consider the following conditions:

1. Is heat capacity of structure high or low?
2. Is structure insulated?
3. Is building exposed to high wind?
4. Is the ventilation or infiltration load high?
5. Is there more glass area than normal?
6. What is the nature of occupancy?
7. What auxiliary heating devices will be in the building?

If the structure has a low heat capacity, is not insulated, has more than normal glass area, or is occupied during the coldest part of the day, the Median of Extremes should probably be selected as the outdoor design temperature. A moderate heat capacity, some internal heat load, and daytime occupancy would indicate the 99% value as a reasonable choice. Massive, institutional buildings with little glass can usually be designed from the 97.5% value.

4.1 Fuel Input Requirements

After completing a heat loss calculation for conventional heating equipment, we must multiply the actual heat loss by at least 1.25 to determine how much natural gas, propane, or fuel oil must be required to heat the building. We are assuming an 80% thermal efficiency. The actual efficiency is closer to 70%. The reason at least 20% of the fuel burned must go up the flue pipe or chimney, as mentioned in Chapter 3, is to keep the water vapor of the combustion process from condensing into a mild acid and returning to the heater or boiler and corroding the equipment.

With gas-fired infra-red heating systems, as you will see later, you will multiply the same heat loss at the most by 0.82 to get the gas input requirements to maintain design conditions.

4.2 Stratification

One of the concepts of actual heat loss that is overlooked is the stratification of heat for both IR heating and conventional unit heater heating. There is an obvious difference between the stratification between unit heaters and infra-red heaters.

4.2.1 Unit heaters

In order for unit heaters to satisfy the 70 °F setting of the thermostat, they must heat the area, in effect, from the ceiling down. This means that when the thermostat, set at 70 °F, is satisfied, the air above the thermostat location is much higher. The heated air coming from the unit heaters expands and, by the law of nature, rises to the ceiling. Temperatures at the ceiling have been measured as high as 130 °F. Since heat loss is a function of the difference in temperature, ΔT, between the inside air and actual temperature of outside air, you can see how great the actual heat loss is at the roof or ceiling.

4.2.2 Gas-fired infra-red heaters

Due to the fact that gas IR heaters do not heat the air directly and there is no need for fans to deliver heat, thermostats can be set at 68 °F to maintain the same comfort conditions. Gas IR does eventually heat the air by conduction, radiation, and transmission. When the heated air reaches the thermostat setting of 68 °F, the heaters shut down. However, the air above the thermostat is basically less than or close to 68 °F all the way to the ceiling. The heat loss, as a result, is only 68 °F and whatever the outside air may be. Actually the ceiling temperature could easily be 120 °F; however, this possibly high temperature would be from the by products of combustion from the unvented gas IR heaters. It's only exhaust gases that create this condition. This could be considered an advantage with the possible melting of any snow on the roof; however, it cannot be designated as a heat loss.

This is another aspect of IR heating that is overlooked, which shows how infra-red is more energy efficient than conventional unit heater equipment.

4.2.3 Condensation

With unvented infra-red heaters, condensation must be addressed and be included in the heat loss calculations since there is always a possibility that a mildew condition or even worse a "rain" condition may occur. However, as will be illustrated later, there are steps in the heat loss calculation that will address this condition. Be sure to include these steps at all times when doing heat loss calculations. A well-designed ventilation system could rectify this condensation problem by exhausting an adequate quantity of the moist indoor air and replacing that volume of air with air from the outside. Even outside air at 10 °F and 100% RH will become reduced to 21% RH at 65 °F. This condition can be determined using the psychrometric chart. Again any ventilation required could benefit using the energy recovery ventilator (ERV) equipment.

4.2.4 Number of heaters

Even with conventional unit heaters, it is necessary to make certain that airflow is uniform throughout the building being heated. With gas IR heaters, it is imperative that you follow the manufacturer's recommendations on minimum distances between heaters as well as the maximum recommended distances from the walls or between rows of IR heaters. The minimum and maximum mounting heights are also very important. In many cases, it may be necessary to change the gas inputs of the heaters in order to get a more uniform distribution of heat to avoid hot spots. This is also another reason to do each wall individually in your calculations. Larger heaters may be required on walls with doors and windows.

4.3 Correction Factors in Design

Correction factors in IR design may be a function of mounting height and ceiling height. There have been cases where it was possible to wall-mount heaters at 12 to 15 feet height with a ceiling height of 65 feet. When this occurs, it may be possible to multiply the heat loss by a factor as low as 0.74 and still maintain design conditions. This is illustrated in Curve B (Figure 4.5) and Figure 4.3.

In some cases, it may be necessary to mount heaters as high as 40 or 50 feet. Some large manufacturing facilities have enormous overhead cranes where it is necessary to be above the crane ways for gas piping to be out of the way. In these jobs you may have to use heaters that are designed with parabolic reflectors to concentrate the IR heat where you want it to be. Should the heaters be above the cranes, it may be necessary to design micro-switches into the controls to shut the heaters off when they are directly above the crane motors to avoid damage to the equipment. Correction factors still apply using the included charts and graphs.

4.3.1 Mounting heights and clearance to combustibles

Manufacturers' recommended mounting heights depend on whether the heater is mounted in a horizontal manner or at an angle from the horizontal.

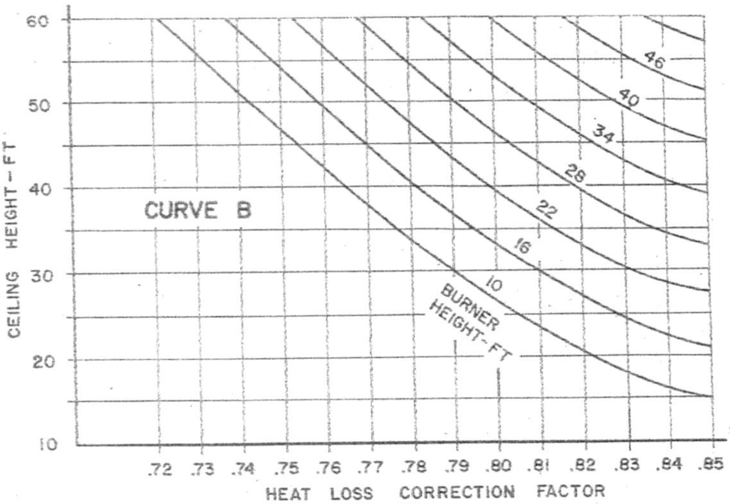

Figure 4.3 Correction factor.

The 30° angle is the recommended maximum angle for the high-intensity IR unit. Any angle greater than 30° could affect the operation of the unit.

The manufacturer's literature for recommended mounting heights usually includes the units with parabolic reflectors. They also list the maximum distance between heaters–usually two times the mounting height. Distances from the wall are also usually listed; however, there should always be enough space between the wall and the heaters to provide clearance for service to the heater.

The maximum distance between heater rows is also provided by the manufacturer. It is usually a function of the mounting height. Since this would mean getting fewer units into the center of the building where there is less heat loss. Outside walls and infiltration (air changes) provide the greatest heat loss in the building. The heater size (gas input) can also be reduced accordingly to meet any spacing requirements.

As for clearance to combustible, it is recommended to check the local fire codes. Usually 10–15 feet above open gasoline cleaning stations is safe, but check with local codes in any case.

4.3.2 Ventilation of byproducts

It is not necessary to vent gas-fired infra-red heaters with flue pipes through the roof as is required with gas unit heaters. By code, they are listed as unvented heaters. However, all buildings heated with gas IR heaters should have approximately 5 square inches of ventilation for each 10,000 Btu of heat input, both inlet and outlet. For example, a standard concrete block (or cinder block) used in building construction is 18" × 8", which could be removed for ventilation requirements. With a louver and screen to keep birds, animals, and insects out, there is 50% of the free area lost in ventilation capability. That reduces the venting-free area to be about 72 square inches, which is enough for about three 50,000 BTUH infra-red heaters even with only natural ventilation.

It is still a good idea to have a thermostat hooked up to a properly sized exhaust fan to remove the byproducts of combustion. There is also the possibility that the building needs an exhaust system for its production or basic use. In many buildings with high ceilings and normal infiltration, adequate ventilation is provided naturally.

Before making any heat loss calculations, it is necessary to make a complete survey of the building, visit the site, or obtain necessary information from the plant engineer if the building is new. Be certain that infra-red is suitable: floors not completely covered with materials; no condensation possibilities; no dangerous vapors such as trichloroethylene, etc.; exhaust requirements of the building; or unusual physical or operational conditions.

4.3.3 Heat transfer

Heat is gained or lost through walls, roofs, windows, doors, concrete slabs, etc., in two basic ways: by transmission through these various materials from air on one side to air on the other side; and by actual leakage of air into the building referred to as infiltration.

ASHRAE method:

The only truly accurate and reliable method of determining a building's heat loss is the ASHRAE method. This system is based on the U-factor, which is an element obtained by careful study and investigation. A U-factor indicates how much heat will transmit through a particular material in BTUs per hour, per square foot, per degree Fahrenheit temperature difference.

4.3.4 Infiltration

Infiltration is defined as the amount of air that will leak into a building due to the differential pressure that exists between the inside and outside of the building. The wind velocity will usually determine pressure difference, unless the building already does have a ventilating system.

Crack method:

The crack method is the most accurate way of determining the heat loss due to normal infiltration. This is explained by the ASHRAE guide and other air conditioning books. Only one-half of the building cracks in filtration should be used because what air enters from one side, it must leave from the other side.

Compute transmission losses using standard ASHRAE methods. No rule of thumb or volume factors. Compute each wall separately. Due to IR units heating the floor, slab losses must be included in overall heat loss calculations. Figure 4.1 will be used as the basis of an IR heat loss calculation. Here we have a building that is 200 feet long by 80 feet wide with a ceiling height of 24 feet. A building ideally suited for gas IR heating. The building materials shown in the problem are typical in many buildings throughout the country: 8 inch concrete block, flat metal walls, steel deck with built-up roofing, and pre-formed 1.5" insulation. Large roll-up doors are also typical. Design parameters will be 0 °F outdoor temperature with an indoor design temperature of 65 °F in an area with 6,848 degree days.

Refer to Figure 4.2 to see the heat loss calculation using U-factors from Table 4.1. For this example the transmission losses amount to 1,070,100 BTUH. The importance of doing each wall individually can now be seen. These individual values will be used to determine the number of heaters required for each wall. We now have to determine the size of the units and the mounting height. In this case, it was determined that a 50,000 BTUH unit angle mounted (20° above horizontal) at 16 feet above the finished floor would be used. Now go to Curve "B" (Figure 4.3). From Curve B, we see a correction factor of 0.83. Curve B (Figure 4.3) was developed to show the effectiveness of the heater mounting height when compared to the overall ceiling height of the building. In this case, it was a 16 feet mounting height to the ceiling height of 24 feet. For buildings that are less than 16 feet high, use a value of 0.85 only.

To determine the minimum amount of air required to dilute the byproducts of combustion, use Curve "C" (Figure 4.4). For this problem, the air requirement is 0.232 (see marked Curve "C"). Multiply 1,070,100 by 0.232 to get 240,770 CFH. The heat required to heat this air to 65 °F is 281,700 BTUH. Our overall heat loss is now 1,070,100 BTUH plus 281,700 BTUH or a total of 1,351,800 BTUH. Use the 0.83 correction factor that we obtained from Curve "B": 0.83 × 1,351,800 BTUH or 1,108,500 BTUH. This is the infra-red input required to heat the building to 65 °F.

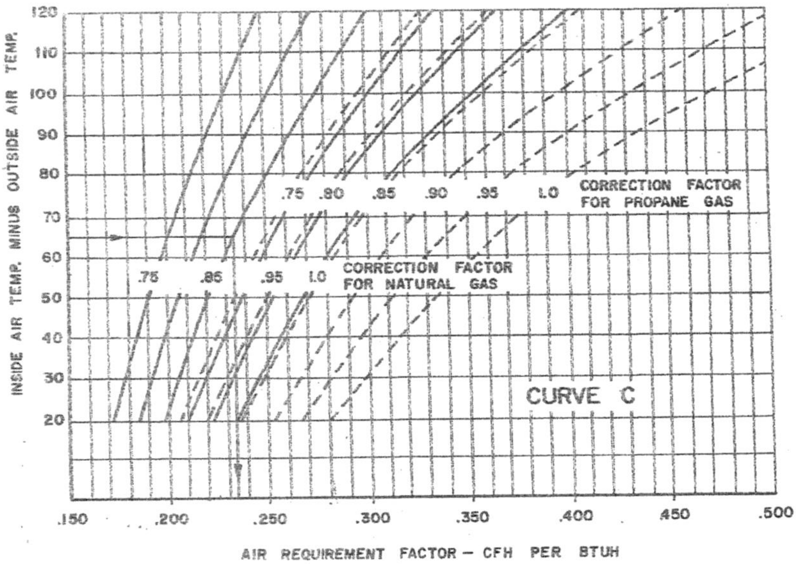

Figure 4.4 Correction Factors.

Now we must check the possibility of condensation. Using the nomograph (Figure 4.5), a line must be drawn between inside temperature minus outside temperature (65 °F) and the U-value of the roof (0.19) to obtain a reading of the inside surface temperature of the roof (62 °F). Since this nomograph was made for inside air temperature of 70 °F, we must subtract 5 °F from the 62 °F to obtain a more realistic 57 °F for the inside of the roof temperature. If the inside room temperature was 70 °F, then the 62 °F would be correct.

In order to determine the dew point of the inside air when the outside air is 0 °F, we must refer to the Dew Point Temperature nomograph (Figure 4.6). Knowing the input of 1,108,500 BTUH and the exhaust of 240 MCFH and connecting these two knowns with a straight line, as shown, we can determine the dew point to be 47 °F. Since the inside surface temperature of the roof is 57 °F and the dew point is lower at 47 °F, condensation will not be a problem. Both nomographs can be used in reverse if more air is required or a different U-factor is needed.

Now that we have resolved the condensation problem, we can determine the number of IR heaters that meet the calculated heat loss. Dividing the heat loss value of 1,108,500 BTUH by the 50,000 BTUH of the selected heater will require 22 heaters.

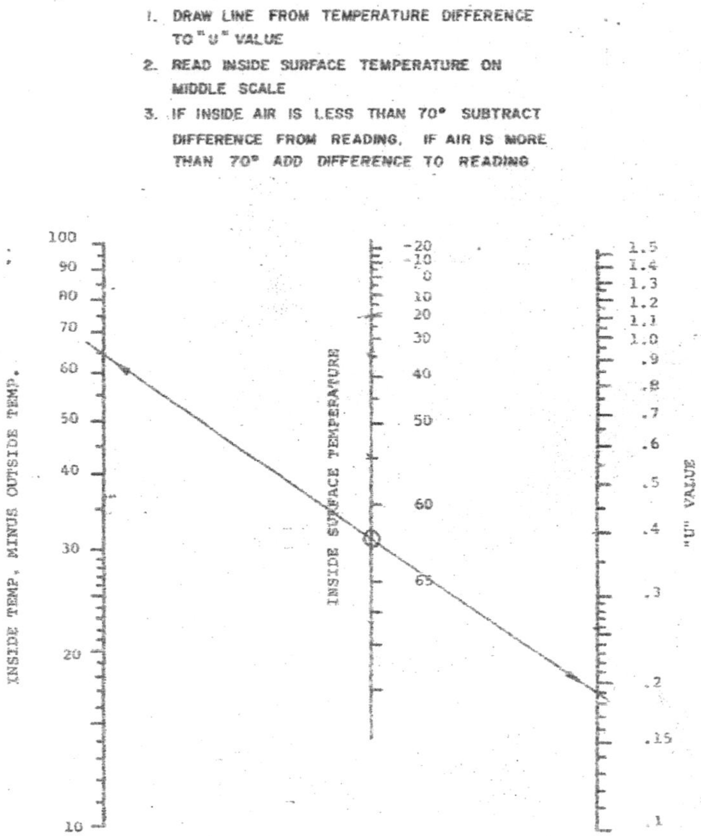

Figure 4.5 Inside surface temperature chart.

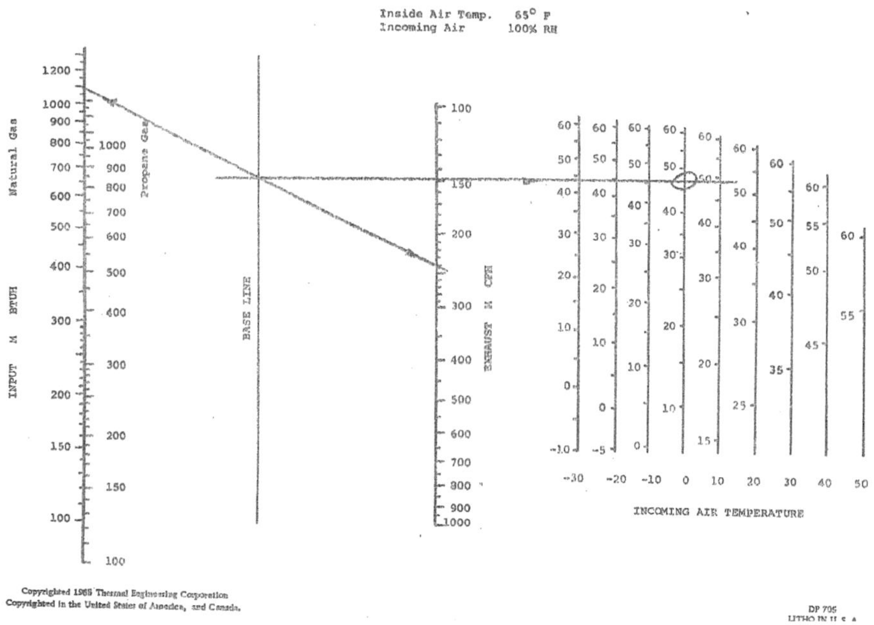

Figure 4.6 Dew point temperature.

5

Heater Layout and Piping Design

There are two ways to install gas-fired IR heaters–either flat-mounted or angle-mounted. This will determine the distance from the wall. Angle-mounted heaters are usually located 3–5 feet from the wall. For flat-mounted 50 MBH heaters, the manufacturers recommend 15–17 feet from the wall. Angle-mounted heaters are mounted between 20° from the horizontal to a maximum of 30°. Most manufacturers of high-intensity heaters recommend a mounting height above the floor for angle mounts of 15–20 feet. For flat-mounts, they recommend anywhere from 17 to 30 feet above the floor. In any case, it is recommended to check with the heater manufacturer's sales representative to determine the proper mounting height. In this example, the heater is angle-mounted at 20° and 16 feet above the floor. See Figure 5.1 for heater layout. Natural gas piping design is the next very important aspect of the overall design.

Gas-fired infra-red heaters have been tested under ANSI code Z83.6 for infra-red heaters. As a result of these tests, IR heaters can be installed using chains, threaded rods, or other methods. See Figure 5.2. Whenever there are overhead cranes, rotating machinery, or other sources of vibration present, it is important that springs or other forms of vibration isolation be installed on all hangers or mounting devices to protect the heaters. In addition, flexible gas lines and flexible electrical wiring should be used to be certain that no vibration is transmitted to the heaters. See Figure 5.1.

5.1 Code Requirements

It is also important to conform to all local building codes or gas utility requirements. In the absence of these local codes or utility requirements, conform to the National Fuel Gas Code, ANSI Z223.1 - 1974. Also of great importance is to make certain that the heaters be mounted level along the long axis of the heater. Should the heater not be installed level and the end of

DOI: 10.1201/9788770046589-7

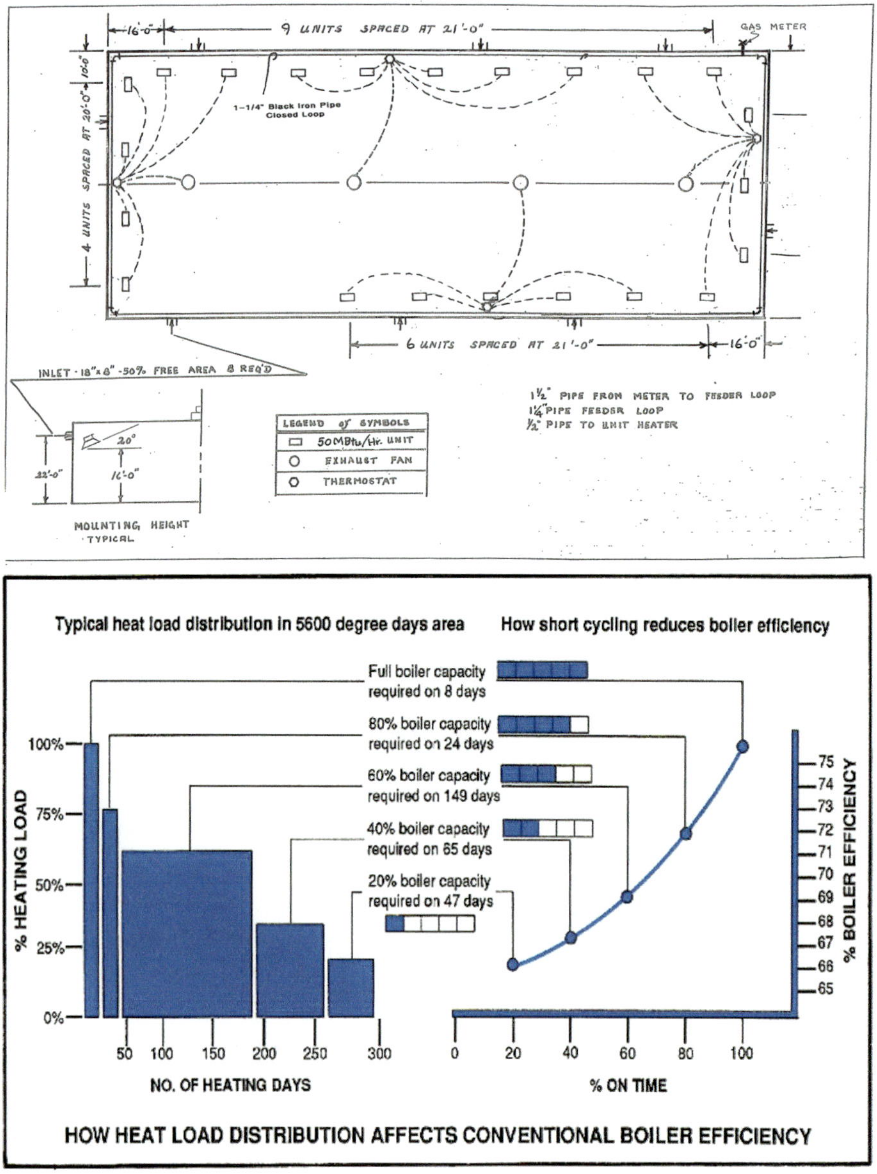

Figure 5.1 Infra-Red Heater Layout.

the heater with controls and pilots on the high side, the operation of the pilot and control can be affected due to the exhaust gases from the main burner damaging these pieces of equipment. See Figure 5.2.

5.2 Manufacturer Requirements

With the high-intensity heaters, it is permissible to tilt the heater to a maximum of 30° from the horizontal. See Figure 5.2.

It is imperative under all conditions to maintain the minimum clearances to any combustibles as indicated on the name plate located on the heater or the manufacturer's specifications on minimum clearance to combustibles. Adequate clearances around air openings into burners should be provided to allow free air flow into the burner venturi. Clearances should also be provided around the heater for accessibility during any necessary cleaning and maintenance.

5.3 Installation

When connecting the gas line to the heater, always install a pipe union in the line adjacent to the pressure regulator. The pipe union used in the gas line should be of the ground joint type or a flanged joint type including a gasket resistant to the action of liquefied petroleum gases. Always place an A.G.A. design certified gas cock in the line upstream from the union. The pipe should then be connected directly into the pressure regulator and should not cause any undo strain on the unit's manifold due to misalignment. See Figures 5.2 and 5.3. Some utilities or local codes require the installation of a condensate trap (drip leg) in the gas piping ahead of each heater to prevent dirt and condensate from entering the heater manifold controls.

5.4 Electric

Wiring to each heater should be installed according to the wiring diagram supplied by the manufacturer of the heater. Each 120 V heater must be electrically grounded in accordance with the National Electric Code, ANSI CI - 1975, when installed if an external electrical source is utilized. Electrical wiring leading to the heater should be installed in such a manner that no undue strain is put on the heater and located away from the hot exhaust areas of the heater to prevent damage to the wiring. An electrical switch should be installed for each heater in the event the heater needs maintenance or needs to be replaced.

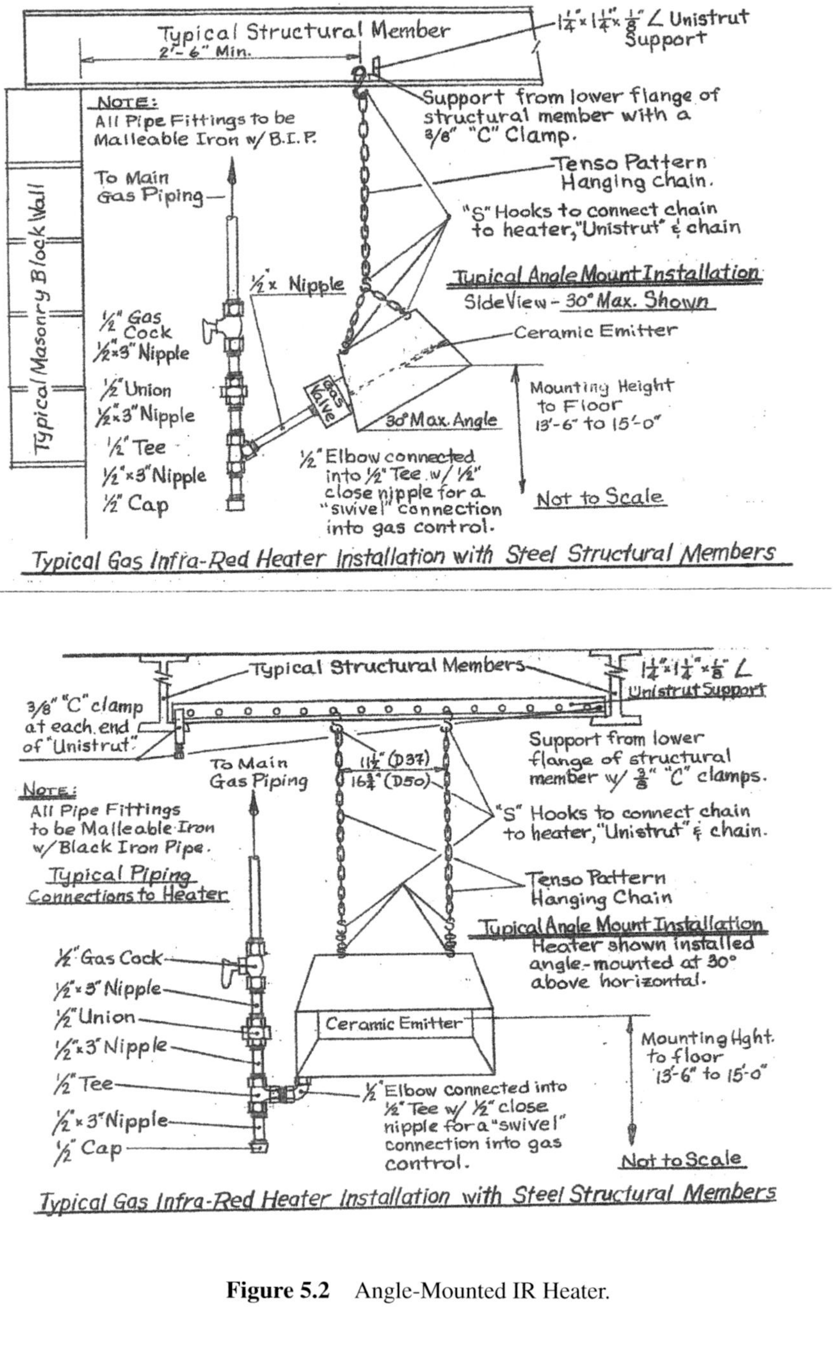

Figure 5.2 Angle-Mounted IR Heater.

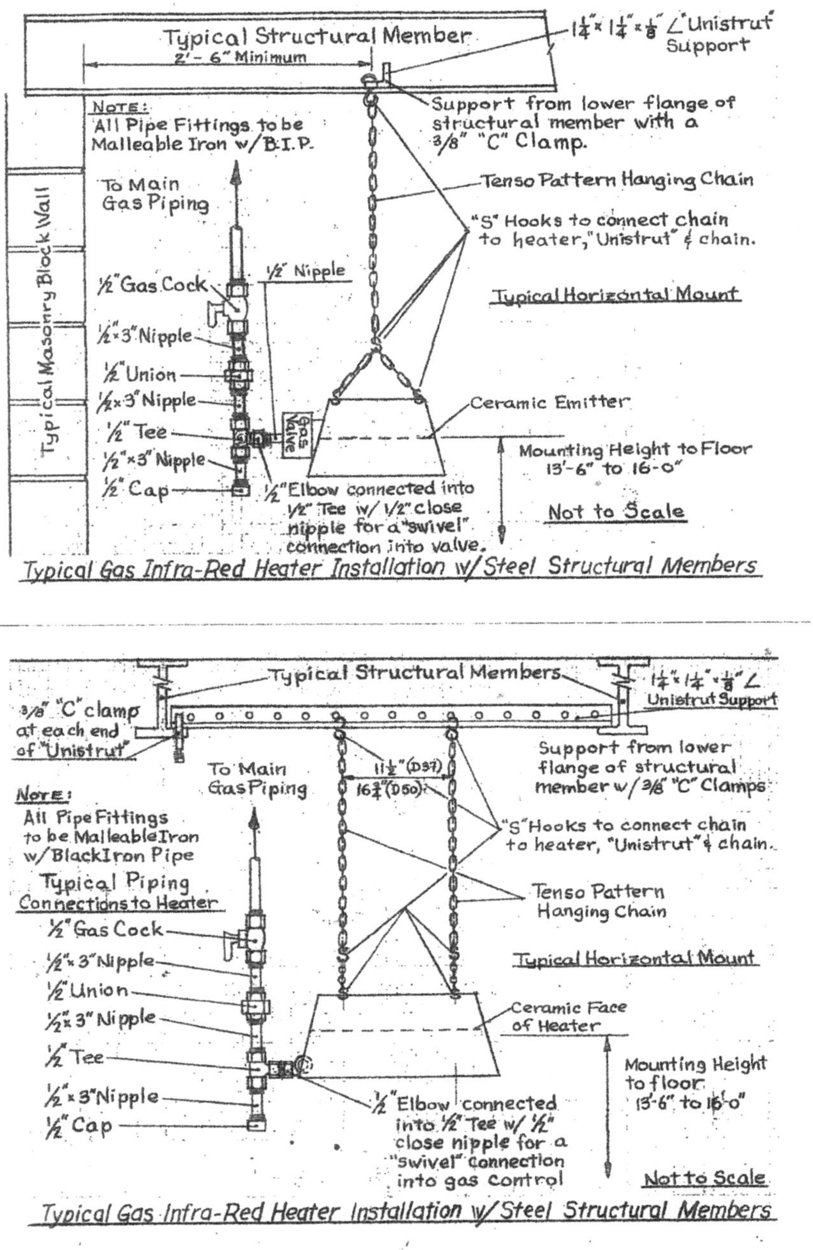

Figure 5.3 Flat-Mounted IR Heater.

5.5 Low-Voltage Systems

Should the heating system be designed for low-voltage (24 V) control, a suitable N.E.C. Class II transformer must be supplied for single or multiple heater installation. The N.E.C. Class II transformer must have enough capacity to provide 25 VA (watts) capacity at the controls of each heater. As with the 120 V heaters, a switch should be installed on each heater for the same reasons.

5.6 Pipe Sizing

The local gas utility companies are the most qualified for designing a gas piping system and their help should be requested. If this service is not available, the following information shown in the pressure drop chart or the table shown and indicated in Chart 5.1 may help in the design of the piping system.

The typical method used for piping design is the trunk line with branches, take offs, to each heater. This method is not recommended for use with high-intensity infra-red heaters. The reason for this is: no matter how over-sized the trunk line is designed, there will still be a slight pressure drop at the end of the trunk line run. With infra-red heaters, this pressure drop, no matter how insignificant it may be, would be noticeable. This shows up in the color or intensity of the last infra-red heater as compared to the infra-red heater color nearer to the gas meter. That is why a "closed loop" piping design is the recommended method. With a "closed loop" piping design, the pressure at each heater, like the air pressure in an automobile tire, is equal all the way around the "closed loop."

Once the IR heaters are located, pipe locations should be shown on the heater plans or separate plan by lines. Tees, crosses, elbows, valves, etc., should also be indicated on the piping layout. Pipe sizes are determined by:

- Cubic feet of gas flow required per hour - CFH
- Length of pipe (including fittings) - FT
- Allowable gas pressure drop - inches of water column

Cubic feet of gas flow - CFH

The volume of gas flow is set by the heating system design and the BTU content of the gas. An IR heater with an input of $50{,}000\,\mathrm{BTUH}$ will require $50\,\mathrm{ft}^3/\mathrm{h}$ of natural gas having a BTU content of $1{,}000\,\mathrm{Btu}/\mathrm{ft}^3$. If propane is

GAS PIPE SIZING

Piping for gas appliances should be of adequate size, and so installed as to provide a supply of gas sufficient to meet the maximum demand without undue loss of pressure between the point of supply (the meter) and the appliance. The size of gas pipe required depends upon the following:

1. Maximum gas consumption to be provided.
2. Length of pipe and number of fittings.
3. Allowable loss in pressure from the outlet of the meter to the appliance.
4. Specific gravity of the gas.

The gas consumption in ft^3/h is obtained by dividing the Btu input rate at which the appliance will be operated by the average Btu heating value per ft^3 of the gas.

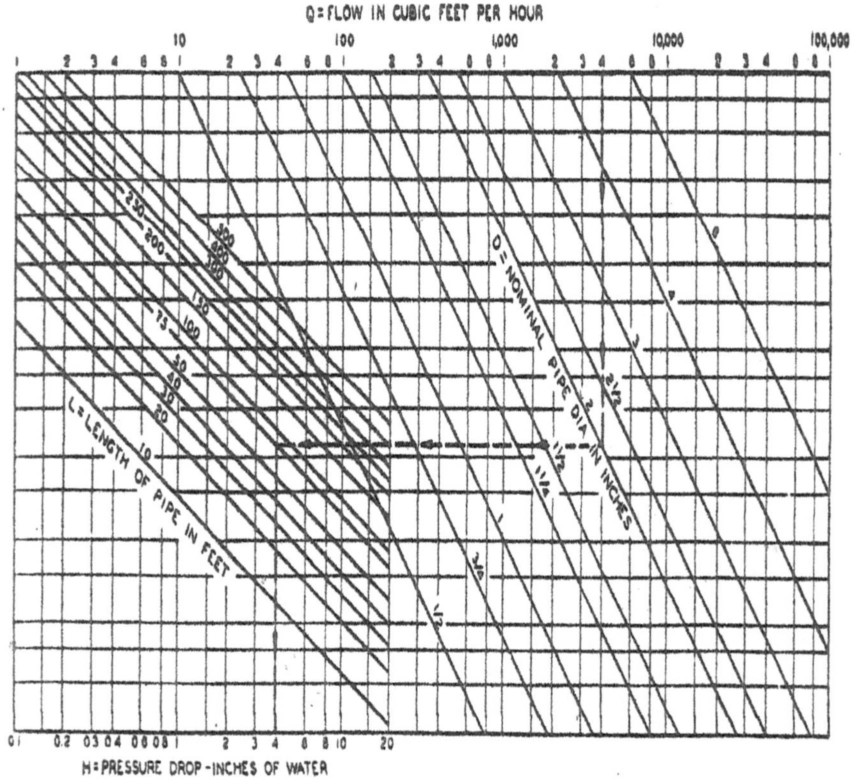

Based on Spitzglass Friction Factor and 0.60 Specific Gravity. To convert for other specific gravity see Table 32.

Fig. 28 Flow vs. Pressure Drop for Low-Pressure Natural Gas

Chart 5.1

the fuel, then a $50{,}000\,\text{BTUH}$ unit will require:

$$\frac{50{,}000\,\text{BTUH}}{2{,}500\,\text{Btu}/\text{ft}^3} = 20\,\text{ft}^3/\text{h}.$$

Length of Pipe - FT

The lengths of pipe between units or branches are also a function of the heating system design. Fittings such as tees, elbows, reducers, unions, etc., add to the pressure drop. Their effect is expressed in equivalent lengths of pipe. Because the pipe size is not immediately known, an estimate must be made. Based on the design shown in Figure 5.1, there are 22 tees to supply gas to each heater, and there are shown 4 elbows and 1 union. Using the information in Chart 5.2 assuming 1" pipe, the equivalent pipe length is 97 inches. Had we assumed a $1\frac{1}{4}$" pipe, the equivalent pipe length is 121.5 inches or about 10 feet. Should the estimate be wrong by a large amount, the estimate will require modification. Experience will make this much simpler.

The volume of gas flow is set by the heating system design and the BTU content of the gas, namely the $1{,}100\,\text{MBH}$ and $1{,}000\,\text{Btu}/\text{ft}^3$ used for the example from Chapter 4.

Inches of water column is the allowable pressure drop through the overall system. This depends on the amount of pressure available at the outlet of the meter. The manifold pressure for most IR heaters is 5.0 inches water column for natural gas and 11.0 inches water column for propane gas. Each IR unit is usually equipped with a pressure regulator set to maintain a 5.0 inches water column in the manifold. Therefore, a pressure of 6.0 inches water column at the meter outlet should be satisfactory as long as the pressure drop through the piping system does not exceed 0.3 inches water column.

If the meter outlet pressure is less than 6.0 inches water column, the IR heater manufacturer should be notified what inlet pressure will be available. Factory adjustment of orifices and pressure regulator will be made to compensate for the lower pressures.

Should the meter outlet pressure be 7.0 inches water column, a pressure drop of 1.0 inch water column will be satisfactory. With a meter outlet pressure of $\frac{1}{2}$" psi, a pressure drop of 3–4 inches water column is usually used. For a meter outlet pressure of 1 psi, a 10 inches water column is usually taken. It can be seen that with higher pressures, smaller size pipe can be used with resultant savings in installation costs.

AGA approvals for IR heaters limit the inlet pressure to 1 psi. If the gas pressure exceeds this amount, two-stage regulation is required. This can be

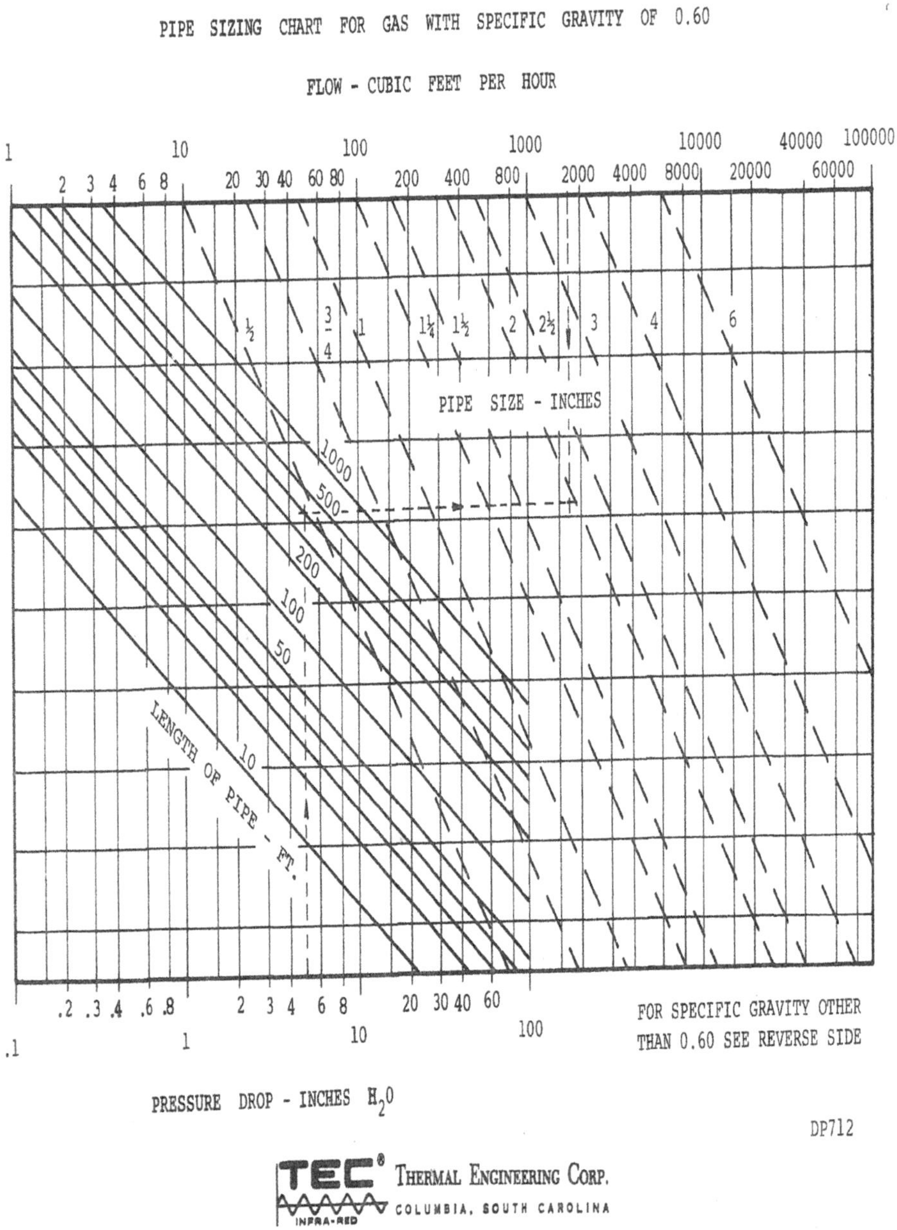
PIPE SIZING CHART FOR GAS WITH SPECIFIC GRAVITY OF 0.60
FLOW - CUBIC FEET PER HOUR
PIPE SIZE - INCHES
LENGTH OF PIPE - FT.
FOR SPECIFIC GRAVITY OTHER THAN 0.60 SEE REVERSE SIDE
PRESSURE DROP - INCHES H_2O
DP712
TEC THERMAL ENGINEERING CORP.
COLUMBIA, SOUTH CAROLINA
INFRA-RED

Chart 5.2

done by an additional regulator for each unit, or by a larger regulator for an entire branch. Economics will be the determining factor.

When the cubic feet of flow, pipe lengths, and allowable pressure drop are determined for any section of pipe, the size can be determined from the pipe sizing: Chart 5.1, Table 5.1, or Chart 5.2.

Assuming the gas service is in the northeast corner of the building, the riser will be 1 $\frac{1}{2}$", which will provide more than 1,100 cfh to the system. The spacing of 21' between heaters is ideal since the standard length of black iron pipe (BIP) is 21'. Starting with the furthest heater from the meter on the west wall will, according to Table 5.1, require 3 lengths of $\frac{3}{4}$" BIP, with a capacity of 150 cfh. The next four lengths will be 1" BIP with a capacity of 285 cfh. Four lengths of $1\frac{1}{4}$" will now be needed providing 590 cfh. The last lengths of BIP will require $1\frac{1}{2}$" in size supplying 890 cfh.

The heaters on the south wall will require three lengths of $\frac{3}{4}$" BIP and two lengths of 1" BIP. In the example design, I would use a closed loop utilizing only $1\frac{1}{4}$" BIP. See Figure 5.3. This may even bring down the overall installed cost using only $1\frac{1}{4}$" fittings as well as only $1\frac{1}{4}$" BIP. The labor cost will definitely be lower.

To further justify the $1\frac{1}{4}$" pipe size, we can again use Chart 5.1. By using an overall 300 feet from the meter to the end of the pipe that includes 10 feet allowance for the $1\frac{1}{4}$ fittings, the "closed loop" is adequate. According to Chart 5.1, there is 170 CFM available at the end of 300 feet. Doing the same for the other feeder shows a total of at least 340 CFM. This is almost 80 CFM more than required.

Using Chart 5.1 to determine capacity available using 1" pipe with the same 0.3 inches of water pressure drop shows 70 CFM with each feeder or a total of 140 CFM available. Based on these data, 1" pipe would actually provide enough gas.

With a closed-loop piping system, all heaters will be operating at the same pressure assuring that all IR heaters will burn at precisely the same surface temperature.

There are other not so obvious reasons that are beneficial to pressure drop in a "closed-loop" system. Since you are introducing the natural gas or propane in effect in both directions, it is building up a positive pressure condition. For this reason, it negates to a certain extent the pressure drop normally associated with pipe fittings. Based on these premises, it is felt that it is possible to get by with 1" pipe for this particular piping design.

Table 5.1 Capacity of gas piping (Ft^3/h)

(At pressure drop of 0.3 in. water. Specific gravity = 0.60)

Length of Pipe Ft	Iron Pipe Size (IPS) Inches									Length of Pipe Ft
	½	¾	1	1¼	1½	2	2½	3	4	
10	132	278	520	1 050	1 600	3 050	4 800	8 500	17 500	10
20	92	190	350	730	1 100	2 100	3 300	5 900	12 000	20
30	73	152	285	590	890	1 650	2 700	4 700	9 700	30
40	63	130	245	500	760	1 450	2 300	4 100	8 300	40
50	56	115	215	440	670	1 270	2 000	3 600	7 400	50
60	50	105	195	400	610	1 150	1 850	3 250	6 800	60
70	46	96	180	370	560	1 050	1 700	3 000	6 200	70
80	43	90	170	503	530	990	1 600	2 800	5 800	80
90	40	84	160	320	490	930	1 500	2 600	5 400	90
100	38	79	150	305	460	870	1 400	2 500	5 100	100
125	34	72	130	275	410	780	1 250	2 200	4 500	125
150	31	64	120	250	380	710	1 130	2 000	4 100	150
175	28	59	110	225	350	650	1 050	1 850	3 800	175
200	26	55	100	210	320	610	980	1 700	3 500	200

From ANSI *Standard Installation of Gas Appliances and Gas Piping*, ANSI Z21.30-1959.

Another thing to consider, although it cannot be included in the actual pipe design, is that it is unlikely that all the heaters will be on at the same time. When the heating season starts, which can be as early as September or October, the thermostat on the coldest wall (north wall) will call for heat. This will then create the beginning of the heat sink in the vicinity of the cold wall. As the heating season progresses, the other heaters will come on as needed expanding the "heat sink." As a rule of thumb, never more than 60%–70% of the units will ever be on at the same time. The heat sink is a major asset of gas infra-red heating.

Again be certain that each IR heater is provided with a $\frac{1}{2}$" gas cock and an electrical switch so that the heater can be shut off in case a heater has to be replaced or to provide maintenance of some kind. Check with the gas supplier to see if they allow a flexible $\frac{1}{2}$" gas appliance connector to be used. The flexible connectors come in various lengths and are made of a flexible material with a stainless steel sheath. Some gas suppliers still insist on black iron pipe with malleable fittings and drip legs.

The heater can be wall mounted with brackets, or with threaded rods with C-clamps, or chain mounted. Make sure that minimum clearances are maintained as recommended by the heater manufacturer. Figure 5.2 shows an infra-red heater mounted at 30° from the horizontal using Tenso Pattern hanging chain and fixed mounting with black iron pipe using malleable fittings. There is a limited flexibility should the unit be hit by a forklift or other piece of equipment. That is why installers of infra-red heaters should get permission to use the $\frac{1}{2}$" flexible appliance connectors. This will provide much greater safety for the heater. Figure 5.3 shows how an infra-red heater can be mounted in a horizontal mode, again from a structural steel member using Tensor Pattern hanging chain in a fixed mounting using black iron pipe and malleable fittings. The only difference between the 30° mounting and the horizontal is the distance from the wall of the heater. The units mounted at 30° are usually 3–6 feet from the wall. Those mounted horizontally are anywhere from 15 to 25 feet from the wall depending on the mounting height. It is best to check with the manufacturer's representative or equipment manufacturer's literature. Figure 5.4 illustrates some other typical suspension hardware depending on whether bar joists, unistruts, steel beams, wooden beams, angle iron, or other supports are used in the building's structure. Threaded rod is often used in the buildings; however, double nut should be used to provide a strong, safe connection as shown in Figure 5.5. Low-intensity tube-type heaters may have its polished aluminum reflectors mounted 45° from the horizontal.

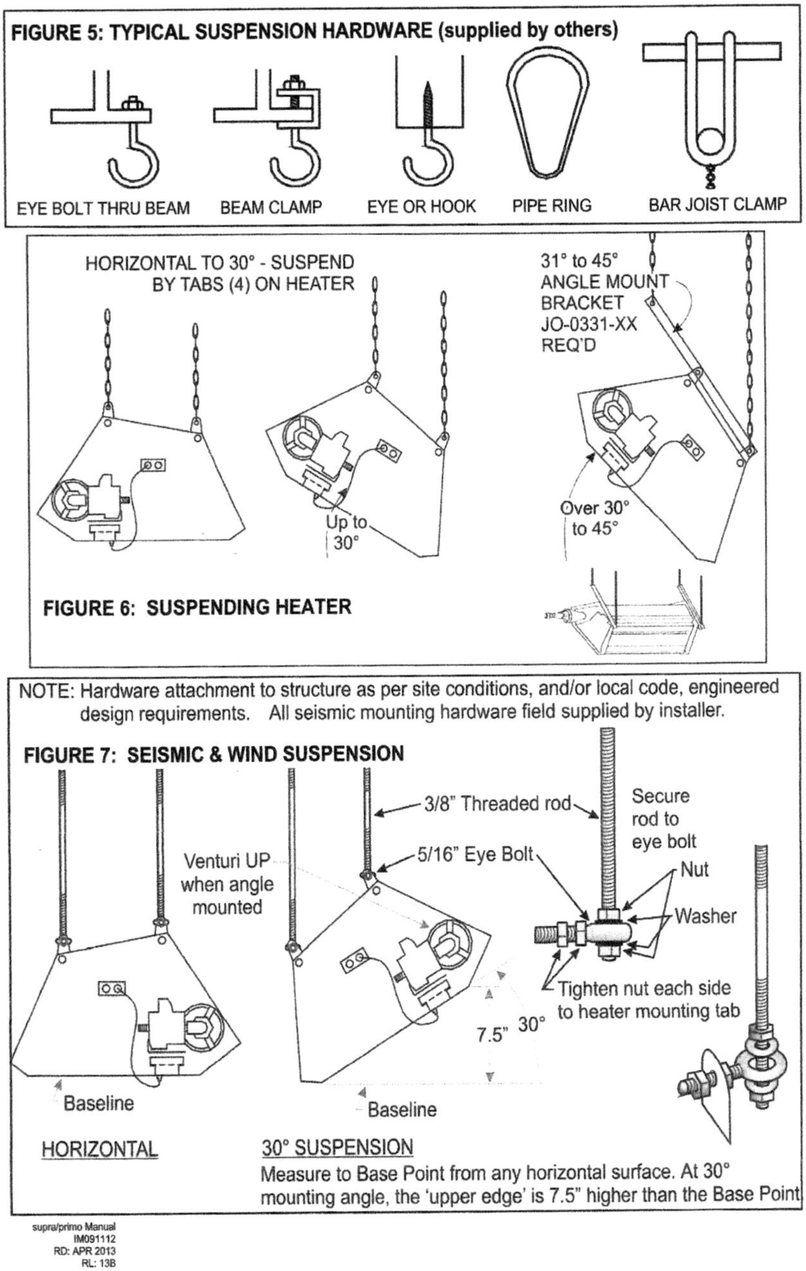

Figure 5.4 IR Heater Hanger Equipment and Hardware.

GAS SUPPLY & CONNECTIONS

Each TEC unit is furnished with a pressure regulator and "A" valve combination. Inlet to this component is 1/2" pipe - female. Most codes specify that a separate shut-off valve and union be provided upstream of the gas connection so that removal of the unit can be made without disruption of the heating system.

The gas connection to the unit should be made with a short length of AGA approved flexible gas line, or with a loop of approved tubing. This will eliminate undue strain on the unit manifold and will facilitate installation.

Gas lines must not be positioned over the heater exhaust slots nor should they be placed in the direct rays of the unit. See sketch below.

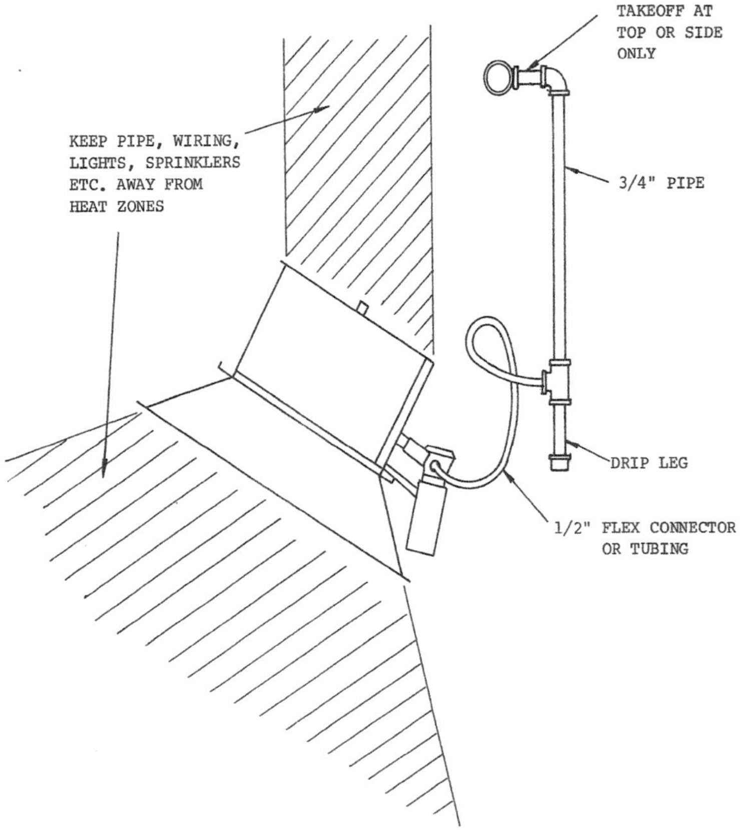

Figure 5.5 Typical Angle-Mount IR Heater.

6

Low-Temperature Infra-Red Design

As mentioned earlier in the book, Roberts-Gordon pioneered gas-fired, low-intensity tube-type IR heaters. This occurred in the late 1950s. Since that time, there have been many other IR heater manufacturers that have taken advantage of this concept.

Radiant heaters designed for low-intensity operation are based on the same principles as high-intensity radiant heaters. The main difference is that the radiating surface is heated to much lower temperatures (500 to about 1,200 °F). At these temperatures, the emitting surface does not glow red, but does emit long-wave heat energy more like the heat from an electric iron or from baseboard radiators.

Low-temperature infra-red, just as high temperature, is a form of radiation that is similar to and performs as light energy. Both light and infra-red are projected from an energy source to an object by wave motion. The main difference between them is the source temperature and the fact that low-temperature infra-red energy cannot be seen by the human eye.

Infra-red energy conforms to the principles of electromagnetics. It forms a small band of the electromagnetic spectrum, which is located between the visible light and the top of the radar and microwave portion of the electronic spectrum. See Figure 6.1. Since infra-red is similar to light energy, it conforms to the same law of optics. Infra-red radiates, as does light energy, in all directions from a point source and travels in a straight line at the speed of light, and can be polarized, focused, or reflected similar to light.

With low-intensity IR heaters, either a power burner or a vacuum-induced burner uses natural gas or propane to push or pull a flame through a 4" diameter steel tube. This tube becomes hot and radiates infra-red energy. These units can easily be vented. However, like the high-intensity IR heater, they are classified as unvented IR heaters. When these units are installed with ventilators, condensation in the building is no longer a problem. However, if they are installed unvented, the condensation possibilities are addressed in the same manner as the high-intensity IR units.

DOI: 10.1201/9788770046589-8

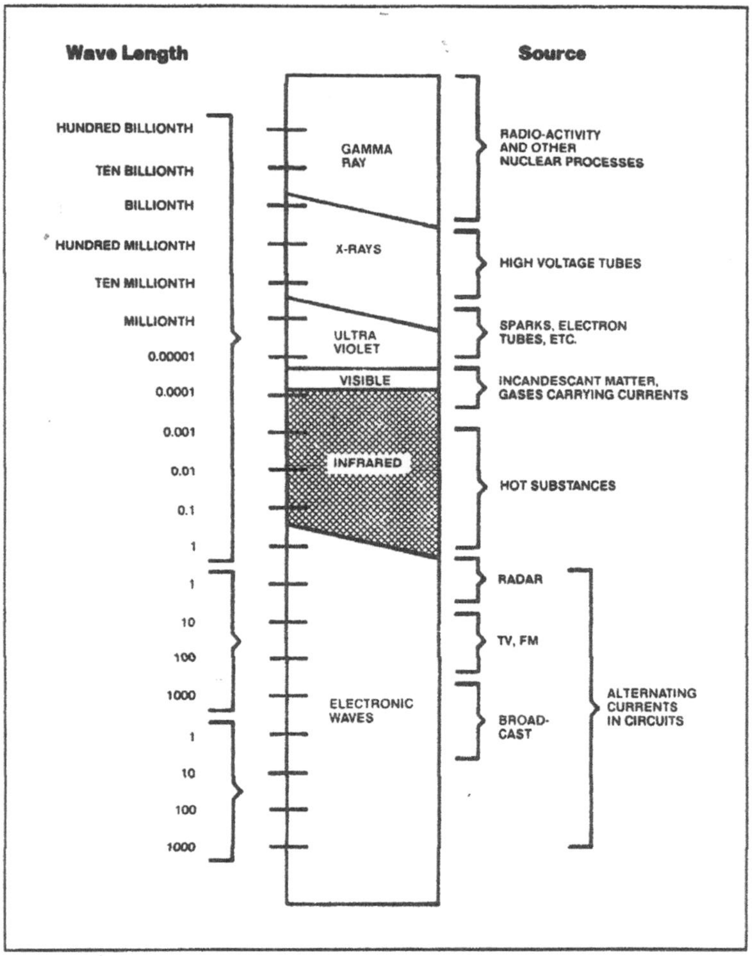

Figure 6.1 Electromagnetic spectrum.

Each type of heater has its own set of operating characteristics and applications, including efficiency characteristics. In thermal (combustion) efficiency, both types of radiant heaters are about the same, that is, their burners extract the same amount of heat from a cubic foot of gas. In essence, gas infra-red heaters have very good system efficiencies for obvious reasons:

1. IR heaters do not heat the air; they heat the objects and occupants by radiation. People can be comfortable at lower air temperatures, which means less heat losses through the building's walls.

2. IR heaters do not require forced air circulation, which also minimizes convection losses through the walls.
3. IR heaters are rated unvented, but must be ventilated at the rate of 3.5–4.5 CFM per 1,000 Btu/h of gas input. This ventilation is usually accomplished with power exhausters installed at the highest point of the roof.

At this point of the basic design, a decision must be made whether to use vented or unvented systems. Things to consider when making this decision are as follows:

1. Presence of chlorinated or fluorinated hydrocarbons
2. Strong possibility of condensation
3. A dusty or dirty atmospheric condition
4. Small volume to be heated
5. Relatively low ceilings
6. Negative building pressures

These conditions usually require low-intensity vented infra-red units.

6.1 Low-Intensity Gas-Fired Vented IR Heaters

There are basically two types of vented IR heaters.

1. Low-intensity gas-fired vented IR units do not require any dilution air. They utilize a power burner firing into usually a 4" outside diameter tubular steel combustion chamber and heat exchanger. They are available in sizes ranging from as small as 10 feet and as long as 70-foot lengths. Gas inputs are from 40,000 to 200,000 BTUH.
2. Low-intensity gas-fired vacuum-vented IR units are also available. They also do not require any dilution air. The unit utilizes a vacuum pump and burner(s) firing again into a 4" outside diameter tubular steel combustion chamber and heat exchanger. They are available in single or multiple burners on a single vacuum pump with inputs that range from 75,000 to 500,000 BTUH gas inputs.

Design procedure for low-intensity IR heaters is the same as for the high-intensity IR units. Compute transmission losses using standard ASHRAE methods. For fresh air requirements for unvented low-intensity IR heaters, use Curve B (Figure 4.3) and Curve C (Figure 4.4). Next we must determine the possibility of condensation using the nomographs displayed on Figure 4.5 and Figure 4.6. The low-intensity IR units are available with input ratings of

40–200 MBH in 5 MBH increments. Tube sizes come from anywhere from 10' to 70' in length. These lengths are accomplished by joining factory pre-assembled 5' and 10' sections. With available accessories, the tubes may be installed in straight, L, U, and compact U configurations. Since combustion is enclosed within the radiant tube, low-intensity may use combustion air from and exhaust directly to the outdoors by simply extending 4" diameter tube through the roof or wall. However, the unit may also take combustion air from and vent exhaust within the building.

Low-intensity IR heaters are suitable for nearly any industrial or commercial building. Applications include relatively low-bay as well as high-bay buildings and warehouses, airplane hangers, factories, automobile garages, etc. Due to its low-intensity, length, and flexible configurations, low-intensity IR units are particularly eligible for lower, well-insulated buildings such as auto dealerships and distribution facilities. Also, since air for combustion and exhaust can be isolated from the building's atmosphere, low-intensity IR units are particularly suitable in buildings with unfriendly environments such as factories under a negative pressure or car wash bays with airborne contaminates as well as body shops.

The standard minimum mounting height of these units varies based on gas input and can range anywhere from 7.5 feet for angle mount to 20 feet for horizontal mount. See the manufacturers' recommendations for specifics.

There are also applications for outdoor covered dining areas when it is not raining or very windy. Small, compact tube heaters are now available for residential garages and workshops. They can be mounted 8 feet above the floor when angle-mounted at 45° from the horizontal. There are many other successful low-temperature infra-red heating system applications that include farm buildings, green houses, fire house garages, gymnasiums, and tennis courts to list just a few. With low-intensity tube-type IR heaters, just use your imagination.

To illustrate the low-intensity design, the building used the high-intensity design system: the layout, the size of the units, and the placement of the units must be considered. For this problem, the mounting remains at 16' above the floor, angle-mounted, 13 feet from the wall. With low-intensity units, the angle of the reflector can be 45° from the horizontal. These IR heaters consist of a burner (either power or vacuum) and radiant tubes with polished aluminum reflectors to which accessories can be added.

See Figure 6.2 for the location and layout of the tubular low-intensity IR heaters.

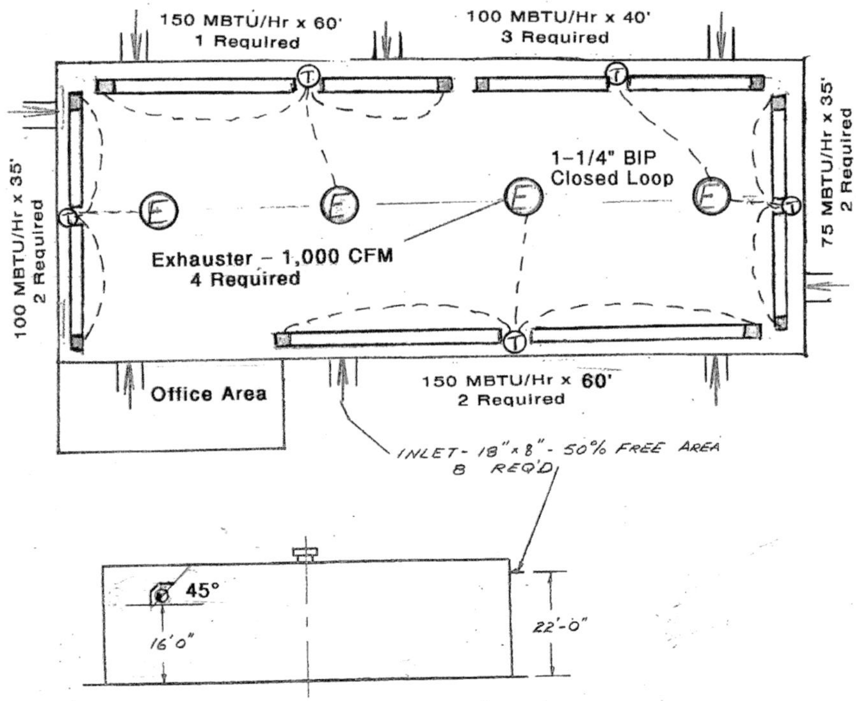

Figure 6.2 Low Temp IR Layout.

It is not necessary to install a closed-loop system for low-intensity tube-type radiant heaters; however, a closed-loop piping system will probably reduce installed cost because the same size pipe and fittings can be used, and also reducing labor costs. There is no way to tell if one heater is at a higher temperature than any other low-intensity unit as you can with the high-intensity units. However, it is recommended to use the closed-loop system to keep installation costs down. Low-intensity units all use either power burners or induced draft fans to generate infra-red and are less susceptible to small gas pressure drops. The problem being used shown on Figure 6.2 would also be safe with a 1" pipe closed-loop system with the slight pressure buildup, inherent with closed loops.

Due to the popularity of the low-temperature IR heaters with their ability to be mounted as low as 8 feet with angle mounting, all the high-temperature IR heater manufacturers mentioned earlier are now manufacturing the tube heaters. Conversely, the reverse is also happening with the low-temperature IR heaters. They all now manufacture high-temperature units. There are many other gas-fired IR heater manufacturers that also make both low- and high-temperature units. The following list is just some of the companies that do manufacture both types of IR heater. However, it is literally impossible to include all:

- AMBIRAD Ltd., Fishers, Indiana 46038
- Detroit Radiant Products Co., Warren, Michigan 48089
- Enerco Group, Inc. (formerly Van Dorn Co., IR Division), Cleveland, OH 44104
- Re-Verber-Ray, Warren, Michigan 48089
- Combustion Research Corp. (Reflect O Ray), Rochester Hills, MI 48309
- Reznor, Mercer, PA 16137

7

Catalytic Infra-Red Heaters

Low-temperature infra-red heaters were developed to provide and supplement the high-temperature units in low bay ceilings that would require energy-efficient IR heating. Gas-fired catalytic infra-red heaters were developed to provide energy-efficient IR heating in hazardous environments. Gas-fired catalytic IR heaters are flameless and explosion proof.

All three types of IR heaters, high intensity, low intensity, and catalytic, emit radiant heat in electromagnetic wave energy in the same manner as light and can be directed, focused, and controlled. Infra-red rays travel at 186,000 miles per second and can be absorbed, reflected, cast shadows, or be transmitted.

The catalytic IR heater consists of a steel or stainless-steel plenum (heater box), gas dispersion plate, an insulation pad, preheat element, catalyst pad, and a wire mesh screen in a bezel. In order to activate, the catalytic heater pad is brought to a required temperature of 250 °F (120 °C) with a low watt density electric heater. When it reached the required temperature, natural gas or propane is introduced into the back of the heater into the perforated aluminum dispersion plate and insulation pad to the catalyst. Flameless combustion producing radiant infra-red heat energy happens when the natural gas (or propane) and oxygen combine within the catalyst pad. Temperatures of about 800 °F (426 °C) are reached on the external face of the catalyst.

Since gas-fired infra-red heaters are flameless and explosion proof, they are a low-cost, efficient, and environmentally safe method of space heating hazardous locations. The catalyst formulation provides a flameless combustion, which lasts as long as there is a continuous supply of natural gas or propane. Catalytic gas IR heaters range in size from 3,500 BTUH to 45,000 BTUH and can be banked to get any Btu range desired.

Due to catalytic IR heater's flameless operation, the units are approved for use in Class 1 – Division 2 Group D hazardous areas by Factory Mutual System insurance and the Canadian Gas Association. They are also approved by the AGA (American Gas Association).

DOI: 10.1201/9788770046589-9

Typical applicates:

- Paint baking systems
- Oil refineries
- Gasoline storage
- PVC pipe bending
- Industrial ovens
- Thermoelectric generation
- Sure-seal systems for yellow jacketed pipelines
- Heating adhesives
- Curing and heating rubber molds
- Curing of plastics
- Rooms/buildings where flammables are stored
- Work areas with hazardous atmospheres

The main reason gas-fired catalytic infra-red qualify for so many varied applications is that they emit the long infra-red waves that offer superior absorption by most materials. This ensures catalytic heaters perform more efficiently than systems that utilize short infra-red wave energy. It also makes the catalytic IR heater ideal for industrial heating, drying, and curing operations.

It should be noted that heat loss is calculated the same as with the other IR heaters. Any process applications will be dealt with in Chapter 11.

BRUEST INDUSTRIES, INC.

Catalytic Heaters

PRODUCE FLAMELESS INFRARED RADIANT HEAT.

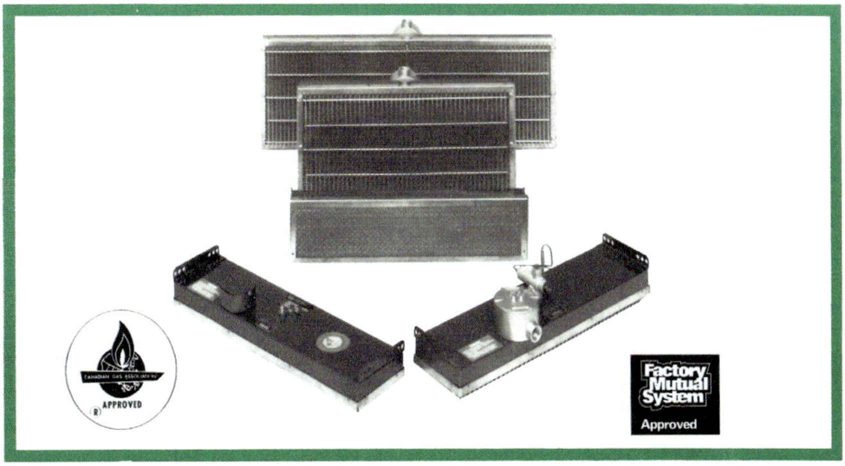

RECTANGULAR MODELS—B6-24 — B12-24 — B12-36 STANDARD MODELS.
— B6-24 — B12-24 — B12-36 FX MODELS — F.M. & C.G.A. APPROVED.

- Bruest Heaters are Extremely Safe.
- Bruest Heaters are Suitable for Use in Some Hazardous Areas.
- Face Temperature Approximately 700° F.
- Factory Mutual System and Canadian Gas Association Approved for Use in Class I, Division 2, Group D, Hazardous Areas.
- Long Life Catalyst Pad - No Moving Parts.
- Over 20 years of Field Proven Reliability.
- Easily Installed in the Field.
- Fuel - Natural or L.P. Gas.
- Additional Sizes Available on Request.

TYPICAL APPLICATIONS

- **Used for Freeze Protection, Space and Pipeline Heating:**

Meter House Heating.
Dog House Heating.
Personnel Heating - Fixed or Portable.
Small Space Heating.
Pipe Line Heating.
Heating Regulating Equipment, i.e.
Regulators
Chokes
Orifice Taps
Meters
Valves

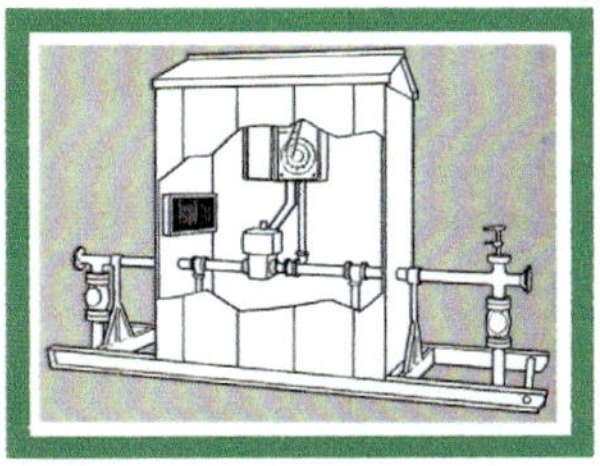

METER HOUSE HEATING

Small Capacity Bruest rectangular heaters are used extensively for heating meter houses and when applied properly, they will prevent freezing in almost any climate. In addition to the extreme safety they provide, Bruest Heaters conserve energy by heating troublesome equipment, frequently replacing larger, more expensive line type heaters.

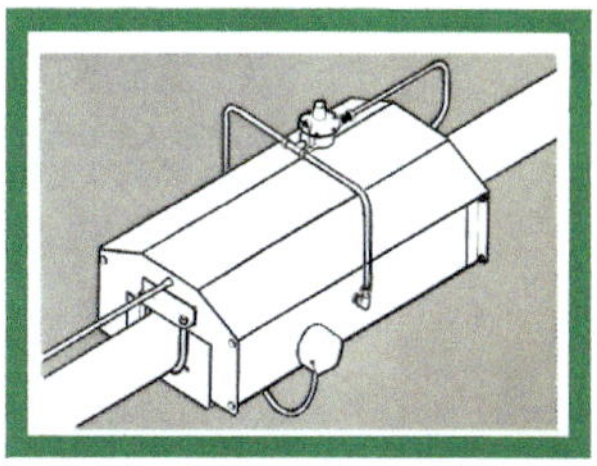

PIPELINE HEATER PACKAGE

Commonly known throughout the industry as the "Twin Pack," the pipeline heater package is ideal for providing the small amount of heat to a gas stream that is necessary where freezing conditions are border line. May be used indoors (specify FX model) or outdoors. Standard models are available in 5,000, 10,000 and 20,000 BTUH input sizes and are complete with low pressure fuel gas regulator, basic piping to regulator and wiring harness.

BRUEST INDUSTRIES, INC. **20th and Sycamore, Independence, Kansas 67301**
(316) 331-0750 • 1-800-835-0557

PRINTED IN U.S.A.

Part II

DOI: 10.1201/9788770046589-10

8

Condensation Considerations

Condensation or the separation of water from air occurs when air is cooled below its dew point temperature. We have all experienced this condition when condensation occurs on the outside surface of a glass containing a cold drink. In my case, a nice dry Martini (shaken not stirred).

Whenever heaters are not vented directly to the outside, they release water vapors into the space they occupy. For every approximately 10,400 Btu of natural gas that is burned, one pound of water vapor is generated. If this air comes in contact with cold surfaces below the dew point, condensation will result.

The inside surface of a roof may be cold when the roof construction has a high U-factor. For instance: with an outside temperature of $0\,^{\circ}$F and an inside temperature of $70\,^{\circ}$F, the inside surface temperature of a bare steel deck ($U = 1.3$) is $15\,^{\circ}$F. If you added insulation to make the U-factor 0.26, the inside surface temperature will be $59\,^{\circ}$F. These values can be determined by using the Inside Surface Temperature Chart illustrated in Figure 4.5. In most cases, dew point temperatures in unvented IR systems with minimum air for dilution will usually be in the $40-50\,^{\circ}$F range. There may be cold spots present, even in roofs with low overall U-values, which are listed below:

1. Single-pane skylights.
2. Compressed batt-type insulation.
3. Thinning out or lack of batt insulation at eave areas or peaks.
4. Insulation without vapor barrier. Moist air passes through insulation and condensation occurs behind the insulation on cold surfaces.
5. Lack of effective seal around skylights, at eaves, etc.
6. False or dropped ceiling not sealed from space above.

In the case of single-pane glass skylights, the U-factor is 1.3, which is highly conducive to condensation. Even with thermal pane glass skylights, there is always the possibility of condensation. The U-factor for double-pane

DOI: 10.1201/9788770046589-11

glass and thermal pane glass is approximately 0.65. The best way to prevent condensation here is to install clear plastic across the skylight opening, well-sealed to prevent air from seeping into the space above the clear plastic.

Should you notice the batt-type insulation has been compressed for any reason, you will find the compressed batting has lost much of its insulating value. When batt-type insulation is compressed, it loses or reduces the air space produced by the batt's original thickness. This condition could reduce its R-value by as much as 50%. Be sure to check to see if the batt-type insulation has been compressed and take this observation into consideration when making any heat loss calculations.

Too often when contractors are installing batt-type insulation in roofs or ceilings, they do not secure the batts and in time gravity takes over and the batts slide down and the peaks become uninsulated. In the case of eaves, it becomes difficult for the contractor to install the insulation and they do what they can to push it into place and damage the batts causing the batts to thin out and lose insulating capability.

Insulation without vapor barriers can cause more than condensation. By having moisture permeate through the insulation, it condenses behind the insulation on the cold surfaces and begins to drip onto the insulation. This causes the insulation to absorb the condensation and lose its insulating characteristics, much in the same way as compressed batt-type insulation. However, there could develop a more insidious condition. There have been conditions when the insulation has held the condensed water vapor and became heavy with the accumulated water (condensation) and came down on the unsuspecting workers and machines with very serious results. Be sure to check if the insulation has a vapor barrier. Most insulation today has vapor barriers.

The lack of an effective seal around skylights may be relating to clear plastic sheets that have been installed across the opening in the ceiling or roof. This is also true around windows, doors, eaves, louvers, and any other openings to the outside.

Other possible areas that may introduce conditions for condensation are false ceilings or dropped ceilings. Here you have to carefully inspect to see that they are sealed from the space above. This could also create conditions like those of insulation without a vapor barrier.

To a high degree, industrial buildings are so constructed that condensation is not usually a problem. However, the building survey, whether it is the actual building or the building plans, must contain an inspection for possible sources of cold spots. Experience has indicated that buildings with low roofs or ceilings must be watched more carefully than buildings with high roofs or ceilings.

Often there are questions of condensation on uninsulated vertical steel walls or glass in outside walls. The Surface Temperature and the Dew Point Temperature Charts show that condensation could occur. However, it has been seen that with IR heaters placed along the perimeter, and spaced not too far apart, condensation will not occur. Exact reasons for this phenomenon are not truly known. A higher mean room temperature and the fact that some radiation does strike the walls may be part of the answer.

Should the building survey indicate that condensation will definitely occur and infra-red is the best solution for building heating, several actions can be taken:

1. Provide more ventilation. This is possible when the increase in ventilation needed is small. Should large amounts be required, more heat, usually in the form of make-up air heaters, is required. Here again is justification for installing the counter flow energy recovery ventilators (ERVs) mentioned earlier in the book.
2. Lower the inside surface temperature by adding some kind of insulation.
3. A combination of the above approaches could also be used to solve the condensation problem.

Whenever the surface temperature of a metal object is below the dew point of the air contacting the surface, there is condensation. Dew point temperature in any space is directly affected by the dew point of the outside air and the number of air changes in the space. Any condensation that occurs on stored products is extremely costly. This is especially true in steel warehouses. Other metals can also be affected, such as galvanized steel, aluminum, and stainless steel.

Fred Prince and Bill Best of Thermal Engineering Corp. (TEC) became interested in solving the condensation problem in the storage of steel in steel warehouses. They solved the problem in several warehouses in the Houston, Texas, area with their Gas Infrared Systems. The design was made by guesstimate, but a good rule of thumb was developed from this experience. It was then found that the rule of thumb would not work where the steel was piled high with little or no space between coil stacks. An engineering approach was needed.

Study and solution of this problem and many others by Prince, Best, and others at TEC during the ensuing years have produced a reasonable approach to the elimination of condensation in storage areas with Gas Infrared Systems.

Fred Prince outlines briefly the causes of condensation and the present art in designing an unvented Gas Infrared Heating System for the prevention of condensation.

Outside dew point fluctuations in steel warehouses throughout the country during the winter months can be obtained from the U.S. Weather Bureau in Rocky Mount, North Carolina. The Bureau keeps records with actual curves, smoothed out, of outside and inside dew points for typical steel warehouses. Using these curves, the temperature that will prevent condensation on the steel can be determined.

In order to prevent condensation, the steel must be heated so that its temperature never falls below the inside dew point. The obvious solution for the above conditions is to keep the warehouse building and steel above 55 °F. However, outside dew point is fickle and has a nasty habit of rising higher than the 55 °F shown above. For most locations, maximum dew point (from ASHRAE data on cooling loads) is about 70 °F.

For a well-insulated warehouse area defined by four walls and a roof, the best and most foolproof condensation prevention system is one that will keep the inside temperature at or above the maximum dew point to be expected in the area. Since maximum dew points happen in the summer with outside temperatures at a much higher level, the above conclusion is safe. Unvented Gas Infrared Systems have practically eliminated any condensation spoilage in such areas. The design is the same as that used for full building heating. Gas infrared's greatest advantages are warm floors with lower fuel costs.

What do you do when a warehouse area in a large building complex does not need heat for its personnel, but only to protect the steel? This condition happens quite often. Always try to convince the owners to activate the IR heaters anytime the temperature falls below 50 °F because as long as the IR heaters are installed, workers will find some way to energize the heaters even when heat is not required for condensation control of the material.

Experience usually shows where condensation will begin. Should the IR heaters be activated at this time, a large amount of heat is needed to keep the steel temperature above the inside dew point. It must be started earlier. To sense the steel temperature and inside dew point temperature and active IR heaters whenever the inside dew point comes within 10 or 15 °F of the steel, the IR heaters will start to run constantly. Now there is a chance to stay ahead of the inside dew point. The rate of steel temperature rise is shown by the higher dew point temperature. Common values are from 1 to 3 °F/h for various parts of the country. Areas near large bodies of water have lower rates of rise but more frequent occurrences of condensation possibilities.

Consider the above as only one severe dew point rise condition. In order to obtain the worst conditions, Weather Bureau data for several years must be analyzed. This has been the practice for many areas throughout the Weather Bureau's upkeep data.

8.1 Ventilation

The primary concern of an HVAC design engineer is to provide a healthy, comfortable indoor environment for the people working in the building. The design must realize that comfort and indoor air quality (IAQ) depends on many factors including not only motor control but addressing the indoor and outdoor pollutants by supplying acceptable air and the removal of unacceptable air. The type of activity the people perform must also be considered.

Ventilation and infiltration are only part of the heating design and the acceptable indoor air quality problem. The HVAC designer and building owner must know and address other related problems as well. They must determine the proper ventilation and infiltration rates to solve the heating problem in a manner to also keep energy usage low and within national, local, and gas utility codes and requirements. These procedures must be approached with care and possibly be under the direction of a professional engineer (PE).

Should there be any concerns with building ventilation and IAQ, it is best to get a copy of ASHRAE Standard 62.1 or 62.2. These standards are reviewed regularly and contain ventilation design and ventilation requirements for commercial and industrial applications.

Any area that contains unvented infra-red heaters, whether it be for full building heating, spot heating, or area heating requires ventilation. Ventilation is best described as the air that is ventilated by gravity or by mechanical means. Air change requirements are usually dictated by the general use of the building or in the case of infra-red heating the possibility of condensation. Buildings that are best suited for infra-red heaters are those that have a medium or high rate of air changes.

Adequate ventilation is required for the combustion and venting of infra-red heaters. Ventilation must be provided to remove the products of combustion from the area above the heaters and inlet air openings below the level of the heaters.

Another requirement for adequate ventilation is the presence of carbon dioxide in the exhaust gases. In the combustion of gases, such as propane or natural gases, carbon dioxide and water vapor are produced. Carbon dioxide is a harmless gas when reduced to safe levels determined by the American Conference of Governmental Industrial Hygienists (ACGIH). The ACGIH-recommended safe level for carbon dioxide is 5,000 PPM (parts per million) and for carbon monoxide 100 PPM. It should be noted that the approval given by the American Gas Association (AGA) for unvented gas-fired infra-red radiant heaters states: "A heater shall produce no carbon monoxide." Carbon monoxide (CO) is not a product of complete combustion but can occur when insufficient combustion air is furnished. Another good reason for adequate ventilation. CO can also be caused where combustion has been arrested before completion, which can happen by providing less than the normal air supply. Incomplete or arrested combustion of natural gas will give off a distinctive odor of aldehydes long before CO concentrations become serious.

8.2 Halogenated Hydrocarbons

Prior to designing a heating system, it is necessary to determine if there are any processes that require or include halogenated hydrocarbons. This includes, but is not limited to, trichloroethylene, perchloroethylene, methylene, chloride, methyl chloroform, and Freon. Some vapors often encountered in industry are decomposed to hazardous or corrosive gases in the presence of high temperature and air. As an example, trichloroethylene, which is used for degreasing, may form phosgene and hydrogen chloride when in the presence of the byproducts of combustion of natural gas. Phosgene is highly toxic, the threshold limit being one (1) PPM. Hydrogen chloride is not as toxic as phosgene, five (5) PPM, but is very corrosive to most metals. It is extremely important to determine these conditions during your initial building survey.

8.3 Power Ventilation

Means for providing supply air is just as important as supplying the means for exhausting air. Dependent upon the construction of the buildings, openings for the air to enter a space may be inherent in the structure itself. This area can be determined accurately by surveying the building and checking for the amount of "crackage" that is present. In many new buildings that are well insulated, have few or no windows, very little area is available for inlet air. In cases of this nature, openings must be provided.

The inlet air area required can be calculated by dividing the exhaust system requirements (CFM) by the allowed inlet velocity of the air in feet per minute (FPM). A velocity of 500 FPM is usually satisfactory. This area should be in small openings located around the perimeter of the building. They should not be located at floor level for they would subject personnel to cold drafts. They should be located near the eave or a few feet below the roof.

Openings on the outside should be louvered, as a weather precaution and should also be screened to prevent birds and bugs of all varieties from entering the building.

Roof-top power ventilators must provide a minimum of four (4) CFM per 1,000 Btu of installed gas input. Check your local code requirements. Ventilation requirements can also be found in the ASHRAE handbook. Some installations use sidewall ventilators where many local codes require 10 CFM per 1,000 Btu. This is because sidewall ventilators are almost always working against outside air movement.

Be careful not to depend on louvers in the end walls to provide inlet air. Should the proposed installation have end wall louvers located high up in the building for summer ventilation, advise the customer to close these louvers during the heating season. Air, like any fluid, will follow the path of least resistance. Therefore, the incoming air will short circuit from the end wall louvers to the rooftop power vents. In any case, mechanical exhaust is recommended in all areas where any other sources of carbon monoxide exist, such as petroleum propane fueled fork lifts. Mechanical exhaust should also have inlet openings of 1.5–3.0 square inches per 1,000 Btu input in order to maintain an inlet air velocity of 200 to 400 FPM.

Power ventilators should always be thermostatically controlled to provide positive ventilation whenever the thermostats call for heat. For small machine shops or welding shops, the power vents should be sized for one or two air changes per hour. The power vents could be wired to by-pass the thermostat and operate all day to remove welding and cutting fumes, and at night they are switched back to operate only when the thermostat(s) call for heat. Normal design procedure is to provide one power ventilator for each thermostatically controlled zone or area. The ventilators should be placed on a maximum of 40′ centers.

Flat roofed buildings should have power ventilators. However, turbine-type gravity vents with a rated CFM may be used if the proper precautions are taken. Avoid using ordinary-type gravity ventilators as back drafts will result. Use extreme caution when using gravity vents with low roof buildings. Ventilation design is an important aspect of gas-fired infra-red heating.

Another method of ventilation control is by means of a humidistat, or other moisture measuring device. As gas is burned, water vapor is formed and this raises the relative humidity of the space.

This method is advantageous where the infiltration and natural ventilation is such that sufficient air is moving through the building a great part of the time. The exhausters then operate only when the natural ventilation is not adequate.

If a humidistat is used to control ventilation, contact the IR equipment engineering personnel for suggested control arrangements.

9

Operating Costs and Comparing Systems

9.1 Design

When designing a gas-fired infra-red system, whether it is with high-intensity units or with low-intensity heaters, we know one thing for sure: it is going to cost more to install than conventional unit heaters. How much more depends on which units are used. The low-intensity infra-red units are going to be much more expensive to install than the high-intensity units.

9.2 Costs

Even though you need more high-intensity IR units than the low-intensity IR heaters for the same heat loss, the heater costs are significantly different. In addition, the labor to install the low-intensity tube heaters is also significant. Tube heaters require at the very least two or more men to mount the units; especially, the very long units. Even the shorter ones need at least two men. Obviously this need affects the overall installation costs.

With the low-intensity tube heaters, you must install 120 V electricity. This requires either BX cable or conduit wiring and transformers to supply 24 V control wiring. Whereas with high-intensity radiant heaters: some cities and communities allow shielded 24 V wiring. Here the 24 V shielded cable can be attached to the piping with nylon wraps going to each heater and its controls. Some municipalities will still require conduits for the 24 V wiring.

However, even with the high-intensity infra-red units, it depends on the size of the units specified. In the example used previously in the book, it was decided to use twenty-two 50 MBH units spaced 21 feet apart. If it was decided to use 60 MBH units spaced 26 feet apart, the installed cost would be reduced due to fewer heaters and the labor costs for each installation. In both cases, we have met the heat loss requirements. The main difference is the overlapping of the infra-red rays. The intensity of the radiation drops off

DOI: 10.1201/9788770046589-12

rapidly as you move further from the heater. As a result, the 50 MBH units spaced 21 feet on centers provide a more uniform heating pattern due to the overlapping of the infra-red radiation. Using 60 MBH units spaced at 26 feet on center will be less uniform and has more "hot spots."

Installation costs could be dramatically reduced had 16–75 MBH units spaced 32 feet apart or 11–100 MBH units spaced 40 feet on centers. Even if the mounting heights were increased from 20 to 25 feet respectively, the hot spots would be more intense and noticeable. This is why, as an infra-red design engineer, you should consult with the manufacturer's representative or the equipment literature to determine the manufacturer's recommended minimum and maximum spaces between the usually high-intensity infra-red heaters. But you still must use a certain amount of discretion. In warmer sectors of the country, it is obvious that you would have to go with smaller heaters with similar spacing to maintain a more uniform distribution of radiation. Likewise in colder areas of the country, you would need larger sized units with controlled spacing in accordance with the heating equipment representative or manufacturer's literature.

It is not really necessary to have even or uniform coverage throughout the building since when using infra-red heating most of the heat loss is through the outside wall. This will require that the major requirements for heaters will be concentrated on these outside walls. As mentioned above, the radiation of the infra-red will overlap to some extent. The following eight tables (Tables 12.1 through 12.8) will give the approximate size of the area covered by an infra-red heater at various mounting heights and gas inputs. The Btu figures given and the size of the rectangle covered are based on the radiation curves of the infra-red burner. This data can be plotted on a curve, but is not practical for use in the field. The areas covered are based on these curves and can be seen in the tables. The Btu intensity at the edge of these areas is one-half of the Btu on the centerline. There will be radiation beyond the areas given, but the radiation intensity drops off very quickly beyond the edge of the rectangle.

These tables are very useful in designing the most effective spacing and mounting height. The tables, when used properly, can be used to eliminate hot spots.

In order to appreciate how effective this overlapping of the infra-red radiation can be, we must recognize that each heater projects its radiant heat in a definite rectangular area. The radiation coming perpendicular from the heater will be at its maximum intensity at the center of this area. However, the intensity drops off rapidly beyond this area as seen in the tables. It varies as the inverse square of the mounting height as in lighting. That is to say that there will be nine times the intensity at 10 feet as there is at 30 feet.

Energy usage in any space is directly related to the heat loss of that space, which is basically the difference between the inside and outside air temperatures. Again, it is extremely important that each wall is calculated individually. Some walls may include more windows and doors. Other walls may have overhead or roll-up doors. Special exhaust fans or required louvers will have an effect on heater placement or the gas input capacity for the infra-red heaters selected for each wall depending on each wall's specific heat loss. This will also determine the number as well as the size of each heater for the overall design. A well-designed gas IR heating system will provide a smaller temperature differential than will a conventional warm air heating system resulting in an enormous potential savings in energy usages. A difference of 10 °F between the two systems can mean a potential savings of as much as 50% in energy costs each year. This will be shown in the following examples.

Assuming that a good IR heating design has been made, the only conditions that will influence annual fuel consumption are the thermostat settings and any internal heat gains. Fuel consumption is also directly related to the degree days (DD) in a certain locality. A degree day is a unit of accumulated departure, based on temperature difference and time. It is used in estimating fuel consumption and specifically nominal heating load of a building in winter. For any one day, the number of degree days of temperature differs between a given base temperature, usually 65 °F (18.3 and 18 °C in Canada) and the mean outside temperature over 24 hours. These DD are also directly related to the average inside temperature maintained. See the DD Appendix Tables.

The formula for estimating annual fuel usage using average degree days is:

$$F = \frac{\mathrm{HL} \times 24 \times \mathrm{DD}}{\mathrm{EFF} \times \Delta T \times \mathrm{HV}},$$

where:

- F = Fuel usage expressed in units of HV (heating value of the fuel)
- HL = Heat loss computed at ΔT
- 24 = Hours per day
- DD = Degree days will vary widely dependent upon the average inside temperature maintained
- EFF = Efficiency of the system

Heating system	Efficiency
Unvented infra-red (gas)	0.92
Vented infra-red (gas)	0.80–0.90
Electric infra-red	1.0
Unit heaters (gas)	0.75
Unit heaters (steam)	0.60
Unit heaters (oil)	0.60

ΔT = Temperature difference between the average inside temperature and the outside design temperature

HV = Heating value of one unit of fuel, usually one of the following:

Unit of fuel	Heating value
MCF natural gas	1,000,000 Btu
Therm of natural gas	100,000 Btu
Gallon propane	91,500 Btu
Gallon #2 fuel oil	140,000 Btu
KWH	3,413 Btu

The portion of the formula $(\text{HL} \times 24)/(\text{EFF} \times \Delta T)$ will be constant for any building because as the inside temperature increases, ΔT and HL also increase proportionately. For this reason, degree days are the governing factor and degree days vary considerably as the inside design temperature varies. Degree days for most cities and bases are listed in the Appendix: SOLARONIC Degree Days. It should be noted that an inside temperature increase of $5 —\text{-}10\,°\text{F}$ higher average air temperature is required with unit heaters to provide the equivalent comfort provided by an IR heating system. This is true because an IR system provides warm floors and equipment with little or no air movement that creates the evaporative cooling effect. Also stratification is lower.

The following example will help in preparing a comparison of fuel usage between warm air systems and IR systems using the building in the illustrated problem:

- Building heat loss – $1{,}351{,}800\,\text{BTUH}$
- Inside temperature – $65\,°\text{F}$
- Outside design temperature – $0\,°\text{F}$
- Degree days for infra-red – Base of 60 - 5498
- Degree days for warm air – Base of 70 - 8403

Warm air gas usage:

- Heat loss:

$$\frac{1{,}351{,}800\,\text{BTUH} \times 24 \times DD}{0.75 \times 70 \times \Delta T} = 1{,}577{,}100\,\text{Btu}$$

- Efficiency = 75%
- HV = MCF = 1,000,000 Btu
- Usage:

$$\frac{1{,}351{,}800\,\text{BTUH} \times 24 \times 8{,}403}{0.75 \times 65 \times 1{,}000{,}000} = 5{,}590\,\text{MCF}.$$

Infra-red gas usage:

- Efficiency = 92%
- HV = MCF
- Usage:

$$\frac{1{,}351{,}800\,\text{BTUH} \times 24 \times 5{,}498}{0.92 \times 60 \times 1{,}000{,}000} = 3{,}230\,\text{MCF}.$$

Assuming natural gas at $1.50/Therm or $15.00/MCF, the cost would be $83,850 for unit heaters and $48,450 for the IR system. This works out to an annual savings of $35,400.

Percent savings over warm air system:

$$\frac{5{,}590\,\text{MCF} - 3{,}230\,\text{MCF}}{5{,}590\,\text{MCF}} = 42.2\%.$$

9.2.1 Programmable thermostats

Additional savings can also be realized by installing programmable thermostats for night and weekend setback temperatures. Another influencing factor is heat gain within the building from lights, equipment, process heat, etc. However these values would be the same for both systems.

To determine the usage with night and weekend setback with the programmable thermostats, you must use the following two equations:

$$U_2 = \frac{\text{HL} \times 24 \times \text{DH} \times \text{DDd}}{\Delta T \times \text{EFF} \times 168 \times \text{HV}}$$

$$U_3 = \frac{\text{HL} \times 24 \times \text{NSH} \times \text{DDN'}}{\Delta T \times \text{EFF} \times 168 \times \text{HV}}$$

where:

- U_2 is the fuel usage during normal working hours
- U_3 is the fuel usage during night setback hours
- DH is the hours per week at normal temperature
- DDd is the degree days at normal temperature
- 168 is the hours in one week
- NSH is the hours/week at setback temperature
- DDN is the degree days at setback temperature

Assuming:

- 10 hours per day at 60 °F - 50 hours
- Degree days at normal temperature - 5498
- Setback temperature - 50 °F
- Hours/week at setback - 118 hours
- Degree day at setback temperature - 3296

Degree days were selected from the degree day tables in the Appendix.

$$F_2 = \frac{1{,}351{,}800 \times 24 \times 50 \times 5{,}498}{70 \times 0.92 \times 168 \times 1{,}000{,}000} = 824\,\text{MCF}$$

$$F_3 = \frac{1{,}351{,}800 \times 24 \times 118 \times 3{,}296}{70 \times 0.92 \times 168 \times 1{,}000{,}000} = 1{,}166\,\text{MCF}.$$

Total usage:

$$F_2 + F_3 = 824\,\text{MCF} + 1{,}166\,\text{MCF} = 1{,}990\,\text{MCF}$$

The total savings using the programmable thermostats would be an additional:

$$3{,}230\,\text{MCF} - 1{,}990\,\text{MCF} = 1{,}240\,\text{MCF}.$$

It can readily be seen that programmable thermostats are a major energy saver along with a well-designed gas infra-red heating system.

9.2.2 Electrical costs

The above savings do not include the savings in electrical energy. With the rising costs of electric rates throughout the country, these costs should not be ignored. In the example used to illustrate the gas costs and savings, at least six unit heaters would have been required. In order for gas- or oil-fired unit heaters to provide heat, they must have heat exchangers. Heat exchangers are always used in indirect-fired space heaters. Their purpose is to provide efficient transfer of heat from the inside surface of the combustion chamber to the air on the outside. In order to have a good heat transfer, they must utilize a series of tubes to provide a large enough surface area to achieve this transfer of heat. In order to accomplish this and move the now heated air into the area being heated requires a fan with an electric requirement of at least $\frac{1}{4}$ horsepower or more. This would amount to approximately 250 watts. In an area similar to Worcester, Massachusetts, this would mean an average 1,200 full load hours. This calculates to 0.25 kW times 1,200 hours or an estimated 300 kW h for each of the six heaters or about 1,800 kW h. At $0.15 per kW h, this is another $270/year for unit heaters. This is a small amount compared to the savings in gas usage; however, it should not be ignored and be pointed out to the prospective user.

By comparison, low-intensity IR units require a 1/25 HP blower motor for each individual unit with a direct spark ignition that includes a 100% safety shut-off. Using the same method of calculations:

$$\frac{1}{25}\text{hp} \times \frac{0.746\,\text{kW}}{\text{hp}} \times \frac{1{,}200\,\text{h}}{\text{year}} \times \frac{\$0.15}{\text{kW h}} = \$5.38 \text{ per heater per year.}$$

Ten low-intensity units would cost $53.80 for electric usage in addition to gas costs. This would be $264.50 in overall electric cost savings.

The electrical costs for the high-intensity IR units is estimated to be less than $1.00/year. A 25 V spark ignition or glow coil requiring less than 15 watts drops off as soon as the pilot proves it has ignited. These additional savings in electricity are relatively small when compared to gas savings but should be considered.

Should the building being heated have a sprinkler system, it should be pointed out to the building owner that should a major power failure occur, with unit heaters there would be no heat unless they had generators to supply electric to the heating system. With high-temperature IR heaters available with self-generating pilots, similar to the pilots in automatic water heaters, there will be heat in strategically placed units to prevent the sprinkler system from freezing. Each of these heaters also has their individual thermostats that would be mounted on the sprinkler main riser. Most power failures occur in the winter on the coldest and "snowiest" days for extended periods of time. Play it safe with a few strategically placed self-generating gas infrared heaters to protect the sprinkler system at a low cost. The self-generating units can only be high-temperature IR heaters.

10

Feasibility Studies and Economics

In heating any building, the cost of the equipment and the installation costs must be taken into consideration. Due to the fact that gas-fired IR heaters will require a greater number of units than a conventional unit heater system. We must be able to account for or justify the higher investment cost due to the fact that gas-fired IR heaters will require a greater number of units than a conventional unit heater system. Historically, the life cycle or useful life of gas or oil unit heaters has been 15 years. This is due to the breaking down of the metal in the heat exchangers and the aging of the fans and fan motors. In the case of high-temperature gas IR heaters, there are no moving parts except for a simple solenoid for gas flow control and ceramic burners that last more than 15 years. Low-temperature units do have fans, but they are small and are not high-volume air movers. Also the tubes are relatively heavy-duty 4 inch diameter in thickness compared to the lightweight metal used in unit heater heat exchangers. This analysis includes the total owning and operating costs. The time value of money as well as escalating energy costs must be included. Equipment life is also an important factor because the life-cycle analysis includes the cost of not only the first cost of the system but also the cost over the life of the system.

There are basically three ways for a company to pay for a new heating system, upgrade of the existing heating system, or overall renovation of the building including the heating system:

1. From existing funds in its treasury
2. Float a bond issue or sell stocks
3. Borrow the required funds from a bank or lending institution

It is unlikely that the company would use money from its profits to install heaters. Usually this money is budgeted to purchase necessary raw materials for its manufacturing of new products. Once money has been budgeted, for whatever reason, it is difficult to borrow from the source.

DOI: 10.1201/9788770046589-13

Bonds are an example of long-term borrowing, like 20-year treasury notes, which, historically, will be about 3% higher than the inflation rate. It is still a decision that must be made by the board of directors, and in some cases by the approval of the stockholders.

The sale of stock is another source of money for the project. Whereas bonds and bank loans are debts that have a maturity date. Stock is an example of ownership of the firm. That leaves a bank loan as the source of money. Most large companies might be able to borrow money at the "prime rate," which is the interest rate that banks charge its best and most sought after customers. Loans to other, less stable companies are charged an interest rate that is higher by anywhere from 0.5% to several percent. Another factor that must be considered, in addition to the financial strength of the borrower and their ability to repay the loan, is that the interest rate will be dependent on the duration of the loan.

The duration of the loan when gas or oil unit heaters are involved is usually the average life of the unit heaters; this could be as little as 5 years or as many as 25 years. The choice is basically that of the owner, which is usually 15 years for the loan.

To compare infra-red heating to conventional unit heaters, it is necessary to translate all cash flow for each system to an equivalent base. The life-cycle cost analysis must take into consideration the "time value" of money. In effect, a dollar in hand today is worth more than one received at some time in the future. This is why a time value must be used on all cash flows in any economic analysis. The cash flow example assumes that cash flows occur at distinct points in time as lump sums and that interest is calculated and is due at distinct points in time. To develop a cash flow model that shows the effect of compounding of interest payments, the cash flow model is developed either as capital recovery or as a uniform series of present worth.

Capital recovery is a uniform series with present value of $1.00. The equation to determine the capital recovery A/P is

$$\frac{A}{P} = \frac{i(1+i)^n}{(1+i)^n - 1},$$

where:

- A is the uniform series of deposits
- P is the present worth
- i is the interest rate
- n is the number of periods that the interest is compounded

The uniform series present worth P/A is basically the reciprocal of capital recovery:

$$\frac{P}{A} = \frac{(1+i)^n - 1}{i(1+i)^n}.$$

Both of these values are tabulated and listed in the tables in the Appendix.

To illustrate how either the capital recovery or the uniform series present worth may be used for an economic evaluation, we will use the example that was used previously: namely the 200 ft × 80 ft × 24 ft ceiling height building.

Because of the relatively high cost of a gas-fired infra-red heating system compared with any gas-fired unit heaters, it is necessary to calculate an economic analysis. This examination must include installation costs, operating costs, and the cost of money to pay for the installations. It will be necessary to use the 22 IR units in the prior example. If we use gas-fired unit heaters, it would require six 300 MBH units to provide the necessary heat input needed to maintain comfort conditions. However, in order to proceed with this economic analysis, it was necessary to arbitrarily use $3,000 as the cost of labor and materials for the gas-fired unit heaters. The $2,000 estimate for the gas IR units was also an educated guess. I am sure that both values are not accurate. Just about any costs for each type of heater would work in this type of analysis.

Six unit heaters installed at $3,000 each come to an estimated $18,000. Whereas 22 IR heaters at $2,000 each, including piping and electrical, come to an estimated $44,000. Gas usage using the basic formula with degree days is:

- Gas-fired unit heaters – 60,580 Therms
- Gas-fired IR heaters – 32,300 Therms

Additional information must be assumed:

- Equipment life – 15 years
- Cost of natural gas – $1.50/Therm
- Cost of borrowing money – 6%

Based on the above information, it is now possible to make an analysis to determine the feasibility of the project.

10.1 Rate of Return Analysis

A simple rate of return (ROR) analysis is one of the most widely used accurate analysis techniques used in the engineering industry. Even though problems sometimes occur in calculating the rate of return, its major advantage overcomes any occasional problems. The major advantage is that it is possible to determine a solitary unit of merit that is easily understood. Consider the above conditions.

Some building managers use the simple payback analysis to determine feasibility. Simple payback is simply the initial investment divided by the estimated annual savings. In this case, the calculation is the installed cost of $44,000 divided by the estimated annual savings of $42,420 ($\$90,870 - -\$48,450$) or a payback of 1.04 years. However, simple payback does not take into account the effect of interest or any energy rate increases. Due to the fact that managers wish to recover their investment costs as soon as possible, the payback method is often used in association with other decision-making concepts.

By using the present worth analysis with the same data as used in the capital return analysis we can also justify the higher initial cost of infra-red heating systems.

Table 10.1 Present worth method.

Value	Unit heaters	Gas IR system
First cost (P)	$18,000	$44,000
Annual operating cost (A)	$83,850	$48,450
USPW (see table)	9.712	9.712
$A \times 9.712$	$814,350	$470,550
Present worth ($P + 9.712A$)	$832,350	$514,550

Again we choose the system with the lowest present worth cost, which is the gas-fired infra-red heating system. The annual cost method is more effective and easier to understand and to show to prospective customers.

Table 10.2 Annual cost method.

Value	Unit heaters	Gas IR system
First cost (P)	$18,000	$44,000
Annual operating cost (A)	$83,850	$48,450
Capital recovery* (CR) (see table)	0.1030	0.1030
$P \times$ CR	$1,850	$4,530
Annual cost ($A + P \times$ CR)	$85,700	$52,980

From this analysis, using the cost of money at 6%, it is quite obvious that IR heaters have the lowest annual cost. Gas infra-red is the "greenest" form of providing comfort heating. The monthly mortgage payment is $\$4{,}530/12 =$ $377.50.

The rate of return:

$$\frac{P}{A} = \frac{\$44{,}000}{\$83{,}850 - \$48{,}450} \approx 1.24 \text{ years.}$$

Percentage rate of savings:

$$\frac{1}{1.24} \approx 80.4\%.$$

It may be a simple calculation of a payback period. A payback period is simply the period of time needed for the energy savings or other benefits to equal the investment in equipment costs and the related installation labor costs. This is a simple definition for payback period. Some companies also consider depreciation of the equipment investment and income taxes; these two are simply called "payback period."

There are times when an industrial or commercial facility has an existing conventional heating system and is complaining about high gas or oil energy costs. The same type of analysis can be utilized. The same data can be used with the exception that there is no installed cost for the unit heaters due to the fact that they are already installed.

10.2 Low-Intensity Infra-Red Heat

The cost to install low-intensity gas-fired infra-red heaters is much higher than for the high-intensity IR units. It is estimated that it costs close to $11,000 per heater. Again, this is only for the sake of the analysis. This is due to higher heater costs and much higher labor and electric installation costs. We must also adjust the annual operating costs due to the fact that the low-intensity IR heaters are not as efficient as the high-intensity units. Assuming an efficiency of only 85%, the annual heating cost would be estimated to be $52,450 for the low-intensity units. Despite these higher costs, an analysis still favors the low-intensity IR heaters. Let us undertake an analysis using the revised heating costs and the higher cost of the IR heaters.

Despite the high installed cost and the higher annual heating cost, it can be seen that gas-fired infra-red is the wiser choice over unit heaters. The estimated annual savings is over $31,000.

The monthly mortgage cost is less than $1,000 at $944.17. The rate of return is now:

$$\frac{P}{A} = \frac{\$110{,}000}{\$83{,}850 - \$52{,}450} \approx 3.5 \text{ years.}$$

Table 10.3 Annual cost method.

Value	Unit heaters	Low-intensity IR
First cost (P)	\$0	\$110,000
Annual operating cost (A)	\$83,850	\$52,450
Capital recovery 6%, 15 years (CR)	0.1030	0.1030
$P \times$ CR	\$0	\$11,330
Annual cost ($A + P \times$ CR)	\$83,850	\$63,780

Percentage rate of savings:

$$\frac{1}{3.5} \approx 28.5\%.$$

There are not many investment opportunities that will provide a return of 28.5%. Had we calculated the annual savings using set-back thermostats, the savings would be significantly greater.

10.2.1 Electric infra-red

Most people when you mention infra-red heating automatically think of electric infra-red. They are used in the electric infra-red lamp by the bathroom shower or their workbench in the garage or basement. There it makes sense, but to heat an entire building, it is very impractical. If one was to heat the building used as an example in this text with a heat loss of 1,351,800 BTUH, it would require almost 400 kW of electric IR heaters. If we use six electric heaters as we did in the above analyses, each heater would require 66 kW capacity, a separate electric feeder from the service panel in at least $\frac{3}{4}$" conduit and heavy capacity electric wire with a ground. Again, each heater also requires an electric driven fan with a separate electric feeder. This would require a small substation to provide a 1,000 A service for just the heating. Considering the additional cost of internal wiring and switch gear, the cost would be prohibitive.

$$\frac{396{,}000\,\text{V A}}{\sqrt{3} \times 240\,\text{V}} \approx 953\,\text{A}.$$

Not a very practical solution for heating the building in question. However, electric infra-red is a fine source of heat for spot heating a small area of a building. It is a far more effective method of providing heat than electric baseboard or electric unit heaters.

11

Gas-Fired Infra-Red Process Heating

This chapter has been primarily written by Bill Best, president and founder of Thermal Engineering Corp. (TEC). There is no one more knowledgeable in gas-fired infra-red process heating than Bill Best. It was mentioned earlier in the text that Thermal Engineering Corp. got out of the infra-red building heating equipment business in 1984 to concentrate on industrial infra-red process heating applications. Bill realized there was greater profit for TEC in process infra-red heating. The following text is based on a talk given by Bill to the AGA Industrial Gas Sales School on June 21, 1966. So it would be safe to say Bill has progressed dramatically since that presentation.

11.1 Gas-Fired Infra-Red Processing

In recent years, it would be appropriate to describe the growth in gas-fired infra-red processing as phenomenal. The possibilities for the use of this equipment are unlimited. Let us discuss for just a moment the features of infra-red process ovens and burner banks.

In process applications using gas infra-red, the energy is transferred from the source to the absorbing part primarily by radiation. The energy transferred is proportionate to the temperature in degrees Rankin to the fourth power and that of the absorber to the fourth power. Since the primary transfer of energy is accomplished by radiation, it is not always necessary that the emitters (radiant burners) be enclosed as is the case in a convection oven.

In a convection oven in order to impart energy or heat to the part, it is necessary to surround the part with air at a sufficient temperature to raise the part to the desired level. Since the energy has to first be transferred to the air, it is important that the air be contained. Since the air is in contact with all parts of the oven, it is further necessary that precautions be taken that an excess amount of energy is not lost through the oven walls. In the design of

DOI: 10.1201/9788770046589-14

a convection oven, it is usually desirable that the enclosure be well insulated to minimize loss to the outside. There are limited cases where it might be required to insulate an infra-red oven, but in general, this is not required. The burners are usually mounted on a steel frame with suitable reflectors between them. Because the reflectors are made from a material that does not readily absorb infrared energy, the loss from the enclosure to the outside is minimized. Also the installation of the conveyor system is simplified because the ends of the oven can usually be left open.

In many applications, no enclosure at all is required, and the burners can be mounted completely in the open. Typical applications of this type are where it is required to raise the temperature of a part as it passes over a conveyor. In many such applications, gas-fired infra-red space heaters provide an economical, satisfactory source of energy. The advantage of infra-red space heaters is that they are independently controlled and the amount of heat can be varied by the number of heaters that are operating adjacent to each other.

11.2 Why Infrared?

Directional: Infra-red energy can be directed. Therefore, it is very useful for heating specific objects or parts of specific objects. It can be focused in the same manner as light.

Ventilation: Where extensive ventilation is required during the heating process, infrared is usually the most efficient type of energy to use. Large amounts of energy are not added to the air; therefore, extensive oven exhaust does not lower the efficiency as it would in a convection system.

Speed: Infrared is a very fast method of heating. This is achieved because of the difference in the energy level of the emitter and the absorber or the part that is being heated. The heating rate of any particular part is generally limited by how fast the part can absorb heat without damage, and not by limitation of the infrared heating equipment. Extremely fast rates can be employed where the process requires the removal of large amounts of water. The part can be exposed to very high concentrations of infrared energy so long as it contains moisture. This feature of an infrared system has tremendous advantages in the textile industry.

Safety: Radiant ovens can use greater volumes of fresh air for the purpose of diluting paint solvents and other vapors, therefore increasing the safety of operation. This feature again results from the fact that the energy is transferred directly from the source to the object and does not need an intervening heat transfer medium such as air.

Adaptability: Infrared burner banks can be very readily adapted to existing materials handling methods such as conveyors. This feature results from the minimum type of surrounding structure that is needed with an infrared system.

Flexibility: One of the more important features of an infrared system is the ability of fast heat-up and cool-down. Most types of gas-fired infra-red burners operate at full temperature within 3 minutes. When the gas supply is interrupted, the cool-down is also very rapid. Since the oven is only operated when needed, further economy in operation can be achieved. The fast cool-down feature of some gas-fired infra-red burners is very beneficial in process drying of textiles or paper.

11.3 Why Gas as a Source of Energy?

Electricity and gaseous fuel are the two basic energy sources for generating infra-red energy for process applications. Depending on location, electricity costs from four to five times the unit energy cost of natural gas. The radiation efficiency (i.e., percentage of total gaseous energy input to the burner converted into infra-red energy) of gas-fired infra-red burners ranges from 30% to 58%. Efficiencies of electrical radiant heaters have been quoted as high as 80%. There are no known sources of any published test data, which indicates efficiency this high. Since there is not a loss due to products of combustion when electricity is used, radiation efficiencies could be higher than those attainable with a gas-fired burner. However, it is obvious when efficiencies are considered along with the basic energy cost that a gas-fired radiant oven can deliver energy to the part to be heated much cheaper than an electrical infra-red oven. The biggest competition so far as electricity is concerned is confusion. Instead of discussing basic energy cost, the electric boys want to talk about flameless heat, cleanliness, etc. The fact remains, however, that economy of operation is dependent upon maintenance cost, component parts replacement cost, and the cost of the energy delivered to the part. They also overlooked the high cost of the additional electric service capacity required. Add to that the cost of the internal electric distribution wiring and switch gear to the oven location. As an example, a 500 MBH infrared oven would require 150 kVA of additional electric capacity. This, in turn, depending on the service voltage, would require 600–1,000 amps of additional service and wiring capacity.

A gas-fired IR oven would only require gas piping to the location, 500 MBH of IR burners, and control wiring. All of these requirements can be installed at a relatively low installed cost.

What are some additional advantages of gas as a primary energy source?

Life expectancy: Long life can be expected from gas-fired infra-red burners. The burners are usually manufactured from high-temperature materials such as ceramics that have almost infinite life in the temperature ranges for which they are designed. Some manufacturers offer a 10-year guarantee on their burners.

Production: In almost every case where a process lends itself to the use of infrared, an increase in production can be assured. In many cases, the infrared equipment can be installed in conjunction with existing equipment to increase production. If productivity depends upon the availability of high radiant intensity in a confined space, gas-fired burners are capable of concentrating large amounts of energy in small spaces. Some types of burners operate at an input of 50–60 thousand $\mathrm{BTUH} / \mathrm{ft}^2$ of burner surface. Under such conditions, the burner may not operate as efficiently as it might otherwise, but where high concentrations of energy are desired, the capability exists.

Safety: When the proper safety features are incorporated in the design of a gas-fired infra-red oven or burner banks, complete operational safety can be assured. Electronic flame, safeguard systems can be incorporated to continuously monitor the flame. In the event of flame failure or lack of ignition, the fuel supply is immediately interrupted.

Installation: In a process where the energy requirement is in excess of 200–300 thousand BTUH, the initial cost of gas-fired infra-red equipment is usually less than that where electricity is employed. Where the input is small and the control is nothing more than an on and off switch for the electrical infrared elements, the initial cost of the electrical system is less. However, if the operating cost is analyzed, the savings resulting from operating the gas-fired infra-red oven can more than make up the difference in initial cost in a very short time. It is important to remember that where inputs are large, the initial cost of gas equipment is almost always less than that of electrical equipment.

The installation cost of gas-fired systems is comparable at the lower inputs and is almost always cheaper with the larger inputs.

Maintenance: Maintenance is certainly a minimum with the operation and use of a gas-fired infra-red oven or burner bank. Maintenance is usually limited to that which is needed on the controls and premix equipment. There is not a continuous requirement for replacing burned out elements.

Control: Gas-fired infra-red burners can be easily controlled. Temperature can be varied by the control of the input, which directly controls the surface temperature. This can usually be accomplished rather inexpensively with manual or motorized valves. Also, the oven can be designed in zones where individual zones can be controlled separately. The control of the individual zones can be simply an on/off control, or it can be on/off in combination with turn down. This design flexibility is a very important advantage of using gas-fired infra-red burners.

Obviously, there are distinct advantages with infrared process equipment; however, we must recognize that it does have limitations. It is not a cure-all, and each application must be investigated to determine if it is a good application for infrared. If none of the advantages cited above can be employed in the application, then the application may be more suited to a competitive heating method.

Applications: The applications are so varied that it would almost be impossible to discuss all of them in a book of this nature. However, it deserves comment on the more important areas where infrared energy has been used with a great deal of success.

Textile industry: The use of infrared for process drying in the textile industry is rapidly expanding. It is being used almost universally where it is necessary to rapidly remove moisture. In many instances, infrared banks have been used in conjunction with existing convection ovens to increase production. In such instances, the infrared portion is usually employed to remove the bulk of the moisture and the convection oven is used for the final drying process.

Plastics: Infrared is being used extensively in the plastic field. It is used to preheat plastics prior to forming and for baking, curing, and other processes. Infrared energy has been used for curing various reinforced parts such as fiberglass boats and prefabricated swimming pool sections.

Paint drying: Probably one of the first uses and more important uses of infrared energy is for paint drying. Typical case history studies show that time for proper curing has been reduced from as much as 16 minutes to 2 minutes.

Water dry off: Infrared energy has been used to remove the water on both dipped and sprayed frit applications and the time required to remove the water for such applications varies from 1.5 to 3 minutes. Infrared energy is used for water dry off of paper, and this has developed into an excellent application with drying cycles ranging from 1 to 7 seconds. Where sheet metal has been treated and it is required to remove moisture, it can usually be removed in less than 1 minute.

Mass heating: Infra-red has been used successfully for heating objects of large masses where it is required to establish and maintain certain temperature levels. Infra-red has been used to bring parts up to a uniform temperature prior to welding and has been used to control the cool-down rate after welding. It has been used for heating molds, sand cores, liquids, and many other applications where heat transfer to a rather large mass is required.

Ceramics: Several years ago, it was necessary to allow ceramic parts to air dry for sometimes as much as two days before they could be fired. Now, because of the application of infrared energy, it is possible to press the ceramic parts and pass them under an infrared bank before they enter the kiln. Infrared energy is being used to remove the proper amount of moisture at the proper rate. This has resulted in less handling and has saved considerable time in this particular industry.

Miscellaneous applications: Infra-red is also being used in synthetic fiber industries, chemical industries, glass industries, and many, many more.

11.4 Oven Design

11.4.1 Estimating energy required

The theoretical considerations involved in the design of an infrared oven are rather complex and could not be adequately covered here. Any attempt to discuss the considerations involved would only create confusion by going into the theory. In the design of an infrared oven, the heat transfer calculations have to be based upon transient conditions. It would be simplified greatly if steady-state conditions existed, but because of the continuous increase in temperature of the part progressing through the oven, different heat transfer conditions exist at almost any point in the oven. Oven design is based not only on theoretical considerations but also practical experience or laboratory test. The preferable method to design is to first conduct tests on the part that is to be heated. Most manufacturers have available lab infrared ovens in which these tests can be conducted. These lab ovens are designed such that any number of the burners required can be operated and the distance between the burners and the part can be varied. Also, the conveyor speed can be varied to determine optimum design conditions. Laboratory tests become even more important when the absorption characteristics of the part to be processed are not known. During the test, a relationship of time-temperature is established. From this relationship, the length of the oven and the number of burners can be determined for any conveyor speed.

Fundamental calculations can be employed to estimate the total input. The heat required to raise the temperature of the part can be determined by the following equation:

$$Q = W \times c \times (T_1 - T_2),$$

where:

- Q is the heat required/h
- W is the weight of load/h
- c is the specific heat of the load
- T_1 is the temperature that parts must reach
- T_2 is the temperature of parts before entering oven

The above equation gives the theoretical amount of energy required to raise any specific part to a given temperature. The loss due to ventilation can be determined from the following equation:

$$Q = \text{CFM} \times 1.1 \times (T_1 - T_2),$$

where:

- CFM is the cubic feet of air per minute through the oven
- T_1 is the temperature of the air leaving the oven
- T_2 is the temperature of the air entering the oven

Where moisture is being removed, the amount of heat required will be the difference between the enthalpy of the saturated liquid as it enters the oven and that of the saturated vapor at the temperature in the oven. In equation form, this is shown as

$$Q = W \times (h_g - h_f),$$

where:

- Q is the BTUH
- W is the pounds of water to be removed/h

- h_g is the enthalpy of saturated vapor at the temperature of the part
- h_f is the enthalpy of the saturated liquid at room temperature

In addition to heat required for the above sources, there is some loss even from an oven employing radiant burners. This loss depends to a great degree upon the efficiency of the burners used. If the radiation efficiency is low, then considerable energy will be transferred into the air and lost either through the enclosure or because of ventilation. Specific information on any individual burner would be required before these losses could be predicted. As you can see, about the most the industrial gas salesman would be required to do is just estimate for the customer the approximate amount of energy required for the oven. In many cases, the variables could be so many that you would not even want to do this. The various manufacturers of infrared ovens can make specific recommendations for you and your customer if you are able to provide him with certain basic data.

11.4.2 Information to obtain during plant survey

1. The amount and type of product that is to be processed in a specified time. Complete details as to shape and size of the product are needed in addition to the weight or mass. If it is possible, samples of the product to be processed should also be obtained.
2. The specific heat of the product and the absorption characteristics if they are known.
3. If the application is a drying process, the amount of water contained in the product is required along with the desired moisture content after the part is processed.
4. Where coatings are being baked or dried, all technical information from the manufacturer of the coatings concerning curing should be obtained. If explosive volatiles are given off, the explosive limits should be known. Also, the amount of volatility from the coating that is given off in relationship to the amount that is applied to the part should be known.
5. The temperature to which the product is to be heated and the desired properties of the product after it is processed. If the temperature desired is critical, the allowable range has to be known.
6. The type of materials handling equipment and its speed that is to be used in conjunction with the oven.
7. Space limitations.
8. Gas pressure and volume that is available.
9. Insurance requirements: FLA or FM.

Any additional information that relates to the process should also be obtained. The more information the manufacturer has, the better recommendations he will be able to make to you and your customer.

11.5 What Does the Future Hold?

The future use of infrared energy for processing is unlimited. The many advantages that infra-red energy offers in processing simply cannot be overlooked by industrial customers. It will be up to us to explain the advantages and show the customers how they can increase production and save money by applying gas-fired infra-red energy to their processes.

Research is now being conducted on new applications of infra-red energy that will create additional markets for manufacturers of the equipment and additional gas loads for the utilities. Work is now going on in such areas as fog dispersal over airport runways, control growth in greenhouses, infrared lasers for very accurate and clean burning of metal. Another promising area for use of infrared energy is the treatment of sewage. So you see, infrared energy does have unlimited possibilities.

12

Spot or Area Heating

There will be times when it is possible that only a small location or area will require heating for one or only a few people. This "spot" or "area" heating cannot be done using unit heaters due to the fact that any heat provided with unit heaters will not stay in the area where it is needed. This heat will rise and migrate to other parts of the building where it is also not needed.

To supply infra-red heating to this small area or "spot" which is contained in an unheated or partially heated large area is called spot or area heating. Infra-red heaters can improve conditions quite substantially where other forms of heat cannot.

The design of a spot heating system is not easy to achieve, at best, due to the many variables involved. In order to devise a reasonable approximate to the design problem, several assumptions must be made in order to avoid tedious time-consuming calculations.

As a result, years ago, Bill Best of Thermal Engineering Corp. (TEC) developed a Surface Heat Loss Chart to address the spot heating requirements needed for design purposes. This chart, Graph 12.1, provides a reasonably accurate estimate of a person's surface losses.

The primary purpose of spot or area heating is to heat the person. The restoration of the heat lost by the surface of an individual is basically his clothing. Clothing surfaces lose heat basically by convection and radiation. Conduction losses of a person's clothing are insignificant.

The graph shown in Graph 12.1 gives the approximate surface heat requirements for varying situations. Values are given for people who are involved in an activity such as light bench work. For other forms of activity, adjustments must be made per the graph notation.

There are two scales shown on the graph. One is for normal clothing (bottom scale), and the other for heavy clothing (top scale). It is assumed that normal clothing consists of: undershirt, undershorts, regular cotton shirt, and cotton trousers. However, heavy clothing is assumed to be double that of normal clothing.

DOI: 10.1201/9788770046589-15

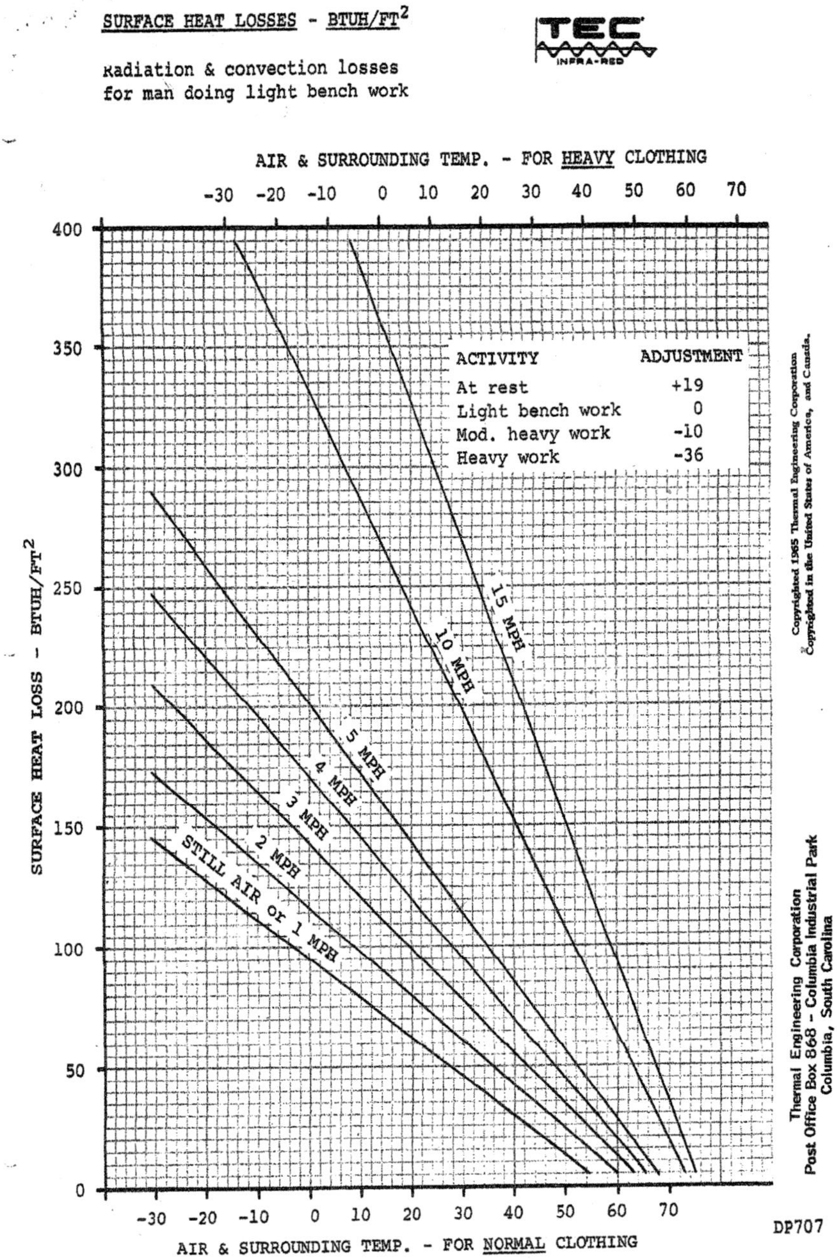

Graph 12.1 Typical heat loss for active individuals.

Another assumption has been made that the temperature of the surrounding objects is equal to the air temperature. This is typically a safe assumption because much of the infra-red energy emitted will pass by the person, be absorbed by the surrounding objects and floor, and thereby rise in temperature.

Now that the surface requirement is established, an equal amount of infra-red energy must be delivered to the surface to offset this loss. The fact that the person is primarily in a vertical plane, the infra-red energy must be delivered to the vertical plane and not to a horizontal plane.

The procedure to prepare a spot heating design is as follows:

1. Determine the design air and surrounding temperature and the clothing to be worn.
2. Determine design air velocity in MPH.
3. Enter top or bottom horizontal scales by air temperature and clothing worn.
4. Follow vertical line up or down until the air velocity line is crossed.
5. At this point, extend a horizontal line to the left and read the surface loss from the vertical scale.
6. Adjust reading (if necessary) per the adjustment table on the Chart.

The amount shown in Step 6 gives the surface loss of the individual in $\mathrm{BTUH/ft^2}$. This is the amount of heat that must be delivered to his body or clothing surface so that he remains in the thermal comfort.

Tables in Figure 12.1 show the infra-red delivery of the IR heater with a parabolic reflector to a vertical surface. Note that the units are mounted at a 30° angle. More infra-red energy is delivered to the vertical surface than if the unit was mounted horizontally. The heaters used in Tables in Figure 12.1 examples are $30\,\mathrm{MBH}$ units with parabolic reflectors. Both tables include multipliers for: 45, 60, 75, and $90\,\mathrm{MBH}$ IR heaters.

The selected heaters should be placed so that the desired delivery is furnished at the mid-point of the person. Wherever possible, the person being heated should be radiated from two sides, front and back. It should be further noted that different units, or combination of units, will provide the same delivery to a certain point or a number of points. The most economical solution, along with the most consistent heat pattern, should be sought and used in the design.

BTUH/FT^2 DELIVERY OF B30 - 30° ANGLE - TO A VERTICAL SURFACE - 1' x 1' - MIDPOINT 3' ABOVE FLOOR

h Ft	d FT 2	4	6	8	10	12	14	16	18	20
7	125	119	81	52	35	25	18	14	12	9
9	44	61	54	42	31	23	18	14	12	9
11	19	28	33	30	25	20	16	13	10	9
13	10	18	21	21	19	17	14	12	10	9
15	5	11	14	15	15	13	12	10	9	8
17	4	7	10	11	11	10	10	9	8	7
19	3	4	6	8	9	9	8	7	6	6
21	2	3	4	6	6	6	6	6	6	5
23	1	3	3	4	5	5	5	5	5	4

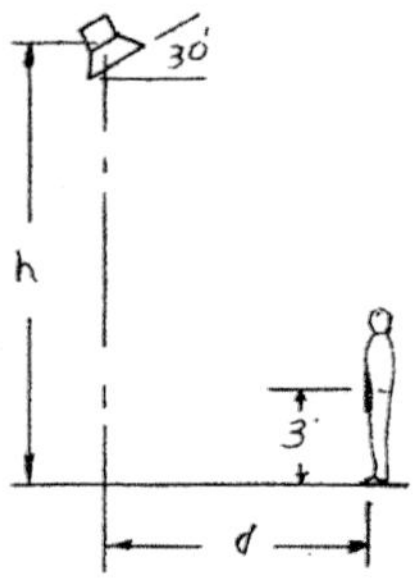

MULTIPLIERS

MODEL NO.	1 UNIT	4 OR MORE UNITS IN LINE** 10' APART	14' APART
B30	1.0	1.68	1.38
B45	1.5*	2.52	2.07
B60	2.0*	3.36	2.76
B75	2.5*	4.20	3.45
B90	3.0*	5.04	4.14

*For section through short axis. Units deliver elliptical pattern. Delivery is greater to sides.

**For average delivery

BTUH/FT^2 DELIVERY OF PB30 - 30° ANGLE - TO A VERTICAL SURFACE - 1' x 1' - MIDPOINT 3' ABOVE FLOOR

h Ft	d FT 2	4	6	8	10	12	14	16	18	20
7	405	242	110	57	27	13	7	3		
9	104	194	109	64	38	25	15	10	6	4
11	37	91	92	61	40	27	19	15	9	7
13	18	48	74	55	39	28	20	15	12	10
15	8	26	46	49	37	27	21	16	12	11
17	6	15	26	39	32	25	21	16	13	10
19	4	9	16	27	30	25	20	14	13	9
21	3	6	10	16	22	20	17	13	12	9
23	2	4	7	12	17	18	15	13	12	9

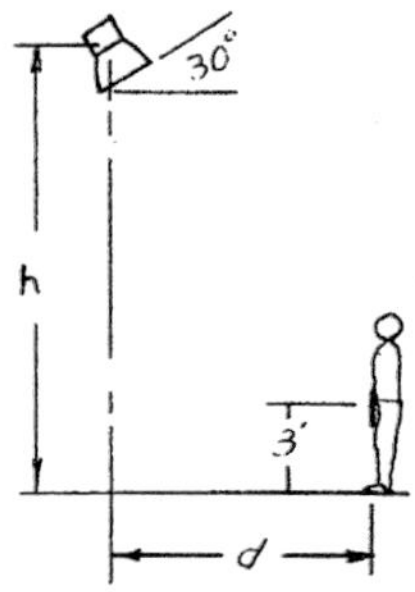

Thermal Engineering Corporation
Post Office Box 868 - Columbia Industrial Park
Columbia, South Carolina

MULTIPLIERS

MODEL NO.	1 UNIT	4 OR MORE UNITS IN LINE** 10' APART	14' APART
PB30	1.0	2.16	1.52
PB45	1.5*	3.24	2.28
PB60	2.0*	4.32	3.04
PB75	2.5*	5.40	3.80
PB90	3.0*	6.48	4.56

*For section through short axis of unit. Units deliver elliptical pattern. Delivery is greater to sides.

**For average delivery

DP708

Figure 12.1 Spot Heating Isolated Individuals with Angle Mounted IR.

If the same requirements existed for an area heating problem and four or more units in line are to be used, 30 MBH units at 14 feet apart, 9 foot high and 8 or more feet away would provide the 51 BTUH. The multiplier for four or more units in line is 1.38; therefore, $51/1.38 = 37$. A position corresponding to 37 must be found in the table in order to deliver the 37 BTUH. The position above approximates this. Again, two rows of units should be used in order that both sides of the individual area are radiated with infra-red energy. The width of the area would be about 17–20 feet.

In some cases, it would be desirable to know the heat delivered to a surface that is parallel to the heater's radiant surface. Each table in Figures 12.2 and 12.3 gives these values but show only one-half of the pattern. The opposite half is identical.

These tables are arranged in the same manner as tables in Figure 12.1, except that the heater is horizontal and the "target" area is also horizontal to provide heat to re-radiate.

If the heater is positioned vertically, then tables in Figures 12.2 and 12.3 can be used to determine the delivery to a vertical surface.

An infra-red spot or area heating job is a compromise, but it can do a job where other methods are completely inadequate.

The Surface Heat Loss graph gives the requirement for one set of conditions. Most of the times, more heat than is required will be furnished as air temperature rises and air velocities decrease. Thermostats that react in the same manner as an absorbing surface are not commercially available. For this reason, it is best to furnish each unit or group of units with an on/off switch and allow the individual to operate the unit or units as his own needs dictate.

An example illustrating the above may further clarify the procedure for determining the "spot" heating design. Assume:

- Air temperature at 30 °F
- Air velocity at 2 MPH
- Activity – moderately heavy
- Clothing – normal

From Graph 12.1, with air at 30 °F and 2 MPH = $61\,\text{BTUH/ft}^2$ are required. From the Graph notation, subtract 10 for moderately heavy work:

$$61\,\text{BTUH/ft}^2 - 10 = 51\,\text{BTUH/ft}^2.$$

BTUH/FT^2 DELIVERY OF B30 - HORIZONTAL - TO HORIZONTAL SURFACE

h Ft	d FT 0	2	4	6	8	10	12	14	16	18	20
4	353	227	89	34	15	7	4	2			
6	157	127	75	40	20	11	6	4	2		
8	88	77	57	36	22	13	8	5	4	3	2
10	57	52	42	31	21	14	10	6	4	3	2
12	42	37	32	25	19	13	10	7	5	4	3
14	29	27	25	20	17	13	10	7	5	4	3
16	22	21	20	17	14	12	9	7	5	4	3
18	17	17	16	14	12	10	8	7	5	4	3
20	14	14	13	12	10	9	7	6	5	3	2

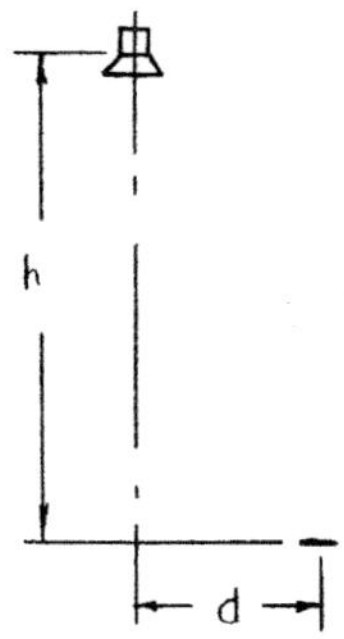

MULTIPLIERS

MODEL NO.	1 UNIT	4 OR MORE UNITS IN LINE** 10' APART	14' APART
B30	1.0	1.68	1.38
B45	1.5*	2.52	2.07
B60	2.0*	3.36	2.76
B75	2.5*	4.20	3.45
B90	3.0*	5.04	4.14

*For section through short axis of unit. Units deliver elliptical pattern. Delivery is greater to sides.

**For average delivery

BTUH/FT^2 DELIVERY OF PB30 - HORIZONTAL - TO HORIZONTAL SURFACE

h FT	d - FT 0	2	4	6	8	10	12	14	16	18	20
4	1288	285	28	5							
6	573	219	76	12	5	2					
8	321	158	76	31	7	2	1				
10	207	123	67	36	15	4	2	1			
12	151	93	56	34	20	8	3	2	1		
14	105	72	49	31	20	12	5	2	1		
16	82	57	42	28	19	13	8	4	2		
18	63	48	36	25	18	12	9	5	3	2	1
20	51	41	30	23	16	12	9	6	4	3	1

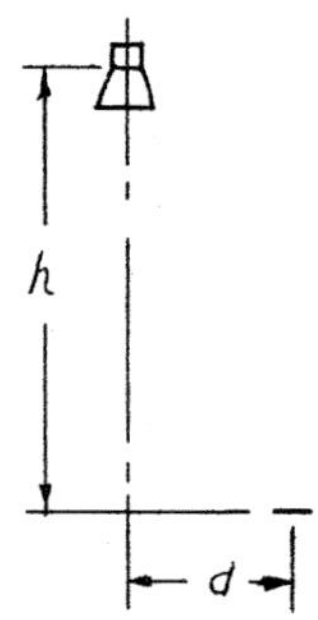

Thermal Engineering Corporation
Post Office Box 868 - Columbia Industrial Park
Columbia, South Carolina

MULTIPLIERS

MODEL NO.	1 UNIT	4 OR MORE UNITS IN LINE** 10' APART	14' APART
PB30	1.0	2.16	1.52
PB45	1.5*	3.24	2.28
PB60	2.0*	4.32	3.04
PB75	2.5*	5.40	3.80
PB90	3.0*	6.48	4.56

*For section through short axis of unit. For delivery through long axis, see **TABLE 5.**

**For average delivery

DP709

Figure 12.2 Spot Heating Isolated Individuals with Horizontal Mounted IR.

BTUH/FT^2 DELIVERY OF PB60 - HORIZONTAL - TO HORIZONTAL SURFACE - SECTION THROUGH LONG AXIS OF UNIT

h	d - FT										
Ft	0	2	4	6	8	10	12	14	16	18	20
4	2576	1388	404	82	10						
6	1146	500	416	180	68	14	4				
8	642	524	348	196	102	48	20	6	2		
10	414	366	268	178	110	64	36	20	8	2	
12	302	260	210	152	106	68	44	28	12	8	4
14	210	194	168	130	98	70	48	34	22	14	8
16	164	150	136	112	86	66	48	36	24	18	12
18	126	124	112	92	76	60	46	36	28	20	14
20	102	100	90	82	66	54	44	36	28	20	16

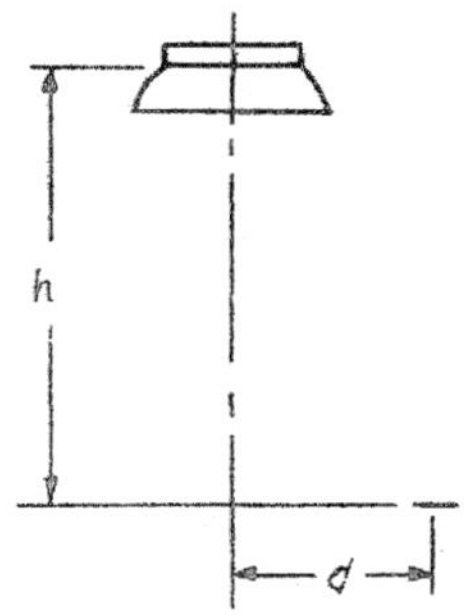

MULTIPLIERS

MODEL NO.	1 UNIT
PB45	0.75
PB60	1.00
PB75	1.25
PB90	1.50

*Above values valid for PB60. Values will vary slightly for other models.

Figure 12.3 Spot Heating Isolated Individuals with Angle Mounted IR.

Table 12.1 indicates 51 BTUH/ft^2 can be delivered to a vertical surface at mid-point of the person if a 30 MBH IR unit with a parabolic reflector is mounted 9 feet above the floor 6 feet from the person being heated. Two units are required. See Figure 12.4 for an illustration.

It should be noted that the delivery at the feet will be less than the 51 MBH/ft^2 design rate and that at the head level it may be greater than 51 MBH/ft^2. It should be further noted that different units or combination of units will produce the same delivery to a certain or a number of points. The most economical solution, along with the most consistent heat pattern, should be sought and used in the design.

12.1 Area Heating

If the same requirements existed for an area heat problem and four or more units in line are to be used, 30 MBH units with parabolic reflectors at 12 feet apart, 9 feet high, and 8 feet plus away would provide the 51 BTUH/ft^2. The multiplier for four or more units in line is 1.38. Therefore, $51/1.38 = 37$ BTUH/ft^2. A position corresponding to 37 BTUH/ft^2 must be found in the table in order to deliver the 37 MBH/ft^2. The position above

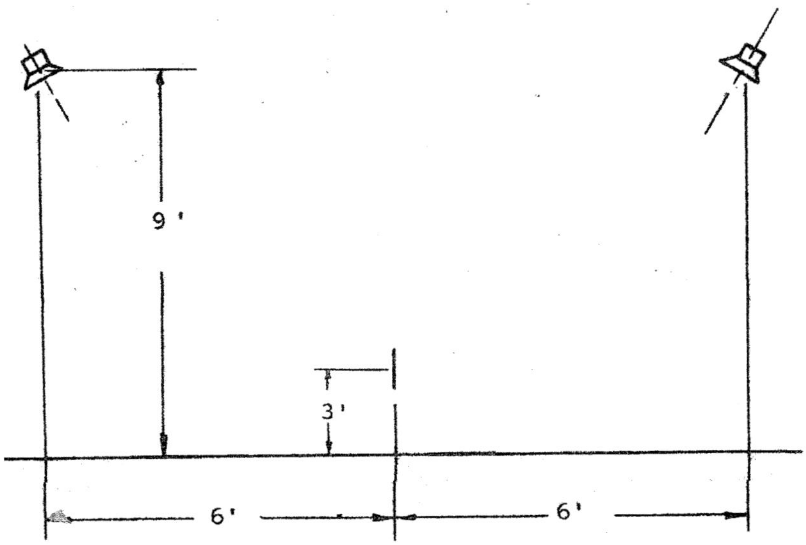

Figure 12.4 Spot Heating Isolated Individuals with Horizontal Mounted IR.

approximates this. Again, two rows of parabolic IR heaters should be used in order that both sides of the individuals in the area are radiated with infra-red energy. The width of the area would be approximately 17–20 feet.

In any case, an infra-red spot or area heating design or installation is a compromise; however, it can do a job where other methods of heating are completely inadequate. See Tables 12.1 through 12.8 for various IR heater gas inputs.

There are many applications for gas infra-red spot or area heating that are only limited by one's imagination. The following applications are only some of the more obvious:

- Snow melting under the marques of hotels, and in some cases motels, to reduce wet shoes from coming into the lobby. Keeping the area from forming ice that patrons could slip and fall. Reduced insurance is also a possibility.
- Golf driving ranges could extend their season for year round operation. Systems could be designed to be coin operated so that the owner need not have to pay for the gas usage.
- Outdoor cafes are great potentials for outdoor dining. Again another way of extending their season. The only drawback would be rain and/or high winds.

Table 12.1 Area Coverage for Mounted Heaters.

Solaronics

AREA COVERAGE

HEATER TYPE: 30 F BTU/HR INPUT: 23, 500 GAS: NAT.

FOR HORIZONTAL MOUNTED HEATERS, A=LENGTH, B=WIDTH OF THE RECTANGLE COVERED

FOR ANGLE MOUNTED HEATERS, SEE THE DIAGRAM BELOW

MTG. HGT.	HORIZONTAL A x B	INTENSITY ℄ - EDGE
10	14. 8 x 13. 9	72 - 36
11	16. 2 x 15. 3	59 - 30
12	17. 6 x 16. 6	50 - 25
13	18. 9 x 18. 0	43 - 22
14	20. 2 x 19. 3	37 - 19
15	21. 6 x 20. 6	32 - 16
16	22. 9 x 22. 0	28 - 14
17	24. 3 x 23. 4	25 - 13
18	25. 6 x 24. 7	22 - 11
19	27. 0 x 26. 1	20 - 10
20	28. 3 x 27. 4	18 - 9
21	29. 7 x 28. 7	16 - 8
22	31. 0 x 30. 1	15 - 7. 5
23	32. 4 x 31. 4	14 - 7
24	33. 7 x 32. 8	12 - 6
25	35. 1 x 34. 2	11 - 5. 5
26	36. 4 x 35. 4	10. 5 - 5. 3
27	37. 8 x 36. 8	10 - 5
28	39. 2 x 38. 2	9 - 4. 5
29	40. 5 x 39. 6	8. 5 - 4. 3
30	41. 8 x 40. 8	8 - 4
31		
32		
33		
34		
35		
36		
37		
38		
39		
40		
41		
42		
43		
44		
45		
46		
47		
48		
49		
50		

MTG. HGT.	ANGLE MOUNTED-30° A x B	'C' DIM.	INTENSITY D	E	F
10	19. 3 x 12. 4	5. 98	50	54	25
11	21. 1 x 13. 6	6. 56	41	45	20
12	22. 9 x 14. 8	7. 14	35	37	18
13	24. 7 x 15. 9	7. 7	29	32	15
14	26. 5 x 17. 1	8. 3	25	27	12
15	28. 3 x 18. 3	8. 86	22	24	11
16	30. 1 x 19. 5	9. 45	20	21	10
17	31. 9 x 20. 7	10. 05	17	19	9
18	33. 6 x 21. 9	10. 6	15	17	8
19	35. 5 x 23. 1	11. 2	14	15	7
20	37. 2 x 24. 3	11. 7	12	13	6
21	39. 1 x 25. 5	12. 3	11	12	5. 5
22	40. 8 x 26. 7	12. 9	10	11	5
23	42. 7 x 27. 8	13. 5	9. 5	10	4. 8
24	44. 5 x 29. 1	14. 1	8. 6	9	4. 3
25	46. 3 x 30. 2	14. 6	8	8. 6	4
26	48. 1 x 31. 5	15. 2	7. 4	8	3. 7
27	49. 8 x 32. 6	15. 8	6. 8	7. 4	3. 4
28	51. 7 x 33. 8	16. 3	6. 4	7	3. 2
29	53. 5 x 35. 0	16. 9	6. 0	6. 4	3
30	55. 2 x 36. 2	17. 5	5. 6	6	2. 8
31					
32					
33					
34					
35					

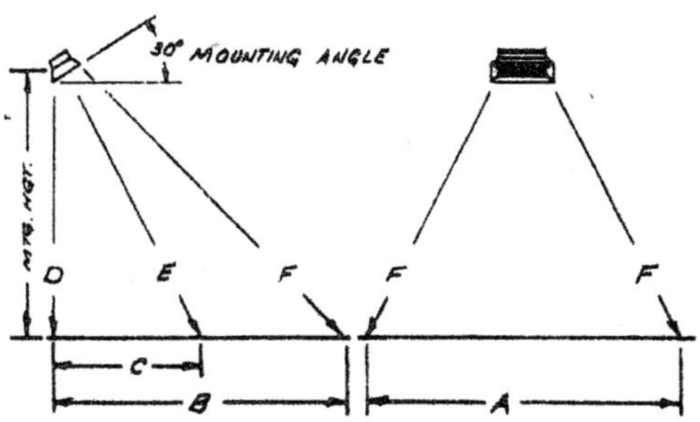

Table 12.2 Area Coverage for Mounted Heaters.

Solaronics

AREA COVERAGE

HEATER TYPE: 45 F BTU/HR INPUT: 37,500 GAS: NAT.

FOR HORIZONTAL MOUNTED HEATERS, A=LENGTH, B=WIDTH OF THE RECTANGLE COVERED

FOR ANGLE MOUNTED HEATERS, SEE THE DIAGRAM BELOW

MTG. HGT.	HORIZONTAL	INTENSITY
	A x B	℄ - EDGE
10	15.4 x 13.9	95 - 48
11	16.8 x 15.3	78 - 39
12	18.2 x 16.6	66 - 33
13	19.6 x 18.0	56 - 28
14	20.9 x 19.3	49 - 25
15	22.2 x 20.6	42 - 21
16	23.6 x 22.0	37 - 19
17	25.0 x 23.4	33 - 17
18	26.3 x 24.7	29 - 15
19	28.6 x 26.1	26 - 13
20	29.0 x 27.4	24 - 12
21	30.4 x 28.7	22 - 11
22	31.7 x 30.1	20 - 10
23	33.0 x 31.4	18 - 9
24	34.4 x 32.8	16 - 8
25	35.8 x 34.2	15 - 7.5
26	37.1 x 35.4	14 - 7
27	38.4 x 36.8	13 - 6.5
28	39.8 x 38.2	12 - 6
29	41.1 x 39.6	11 - 5.5
30	42.5 x 40.8	10.6 x 5.3
31		
32		
33		
34		
35		
36		
37		
38		
39		
40		
41		
42		
43		
44		
45		
46		
47		
48		
49		
50		

MTG. HGT.	ANGLE MOUNTED 30°		INTENSITY		
	A x B	"C" DIM.	D	E	F
10	22.8 x 12.4	5.98	66	71	33
11	24.8 x 13.6	6.56	54	58	27
12	27.0 x 14.8	7.14	46	49	23
13	29.0 x 15.9	7.7	39	42	20
14	31.2 x 17.1	8.3	34	36	17
15	33.2 x 18.3	8.86	29	31	15
16	35.3 x 19.5	9.45	26	28	13
17	37.3 x 20.7	10.05	23	24	12
18	39.4 x 21.9	10.6	20	22	10
19	41.5 x 23.1	11.2	18	20	9
20	43.6 x 24.3	11.7	16	18	8
21	45.6 x 25.5	12.3	15	16	7.5
22	47.7 x 26.7	12.9	14	15	7
23	49.8 x 27.8	13.5	12	13	6
24	51.9 x 29.1	14.1	11	12	5.5
25	53.9 x 30.2	14.6	10.6	11	5.3
26	56.0 x 31.5	15.2	9.8	10	4.9
27	58.0 x 32.6	15.8	9	9.5	4.5
28	60.2 x 33.8	16.3	8.4	9	4.2
29	62.3 x 35.0	16.9	7.8	8.5	3.9
30	64.4 x 36.2	17.5	7.3	8	3.6
31					
32					
33					
34					
35					

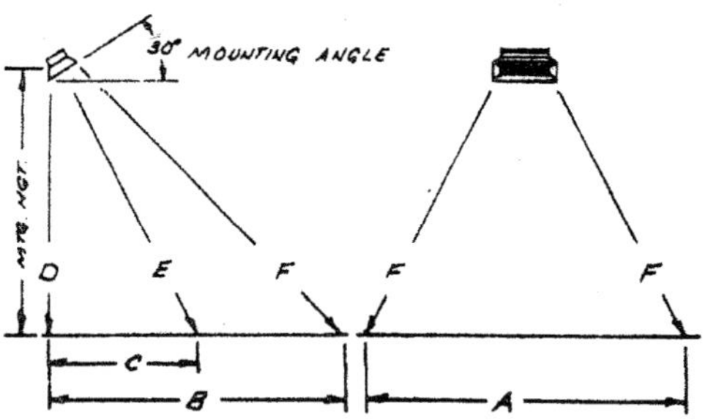

Table 12.3 Area Coverage for Mounted Heaters.

Solaronics

AREA COVERAGE

HEATER TYPE: 60 D-F BTU/HR INPUT: 50,000 GAS: NAT.

FOR HORIZONTAL MOUNTED HEATERS, A=LENGTH, B=WIDTH OF THE RECTANGLE COVERED

FOR ANGLE MOUNTED HEATERS, SEE THE DIAGRAM BELOW

MTG. HGT.	HORIZONTAL A x B	INTENSITY ℄ - EDGE
10	15.4 x 14.1	125 - 63
11	16.8 x 15.4	102 - 51
12	18.2 x 16.8	86 - 43
13	19.6 x 18.2	74 - 37
14	20.9 x 19.5	64 - 32
15	22.2 x 20.8	55 - 27
16	23.6 x 22.2	49 - 24
17	25.0 x 23.5	43 - 21
18	26.3 x 24.8	38 - 19
19	28.6 x 26.2	34 - 17
20	29.0 x 28.6	31 - 15
21	30.4 x 28.9	28 - 14
22	31.7 x 30.2	26 - 13
23	33.0 x 31.6	24 - 12
24	34.4 x 33.0	22 - 11
25	35.8 x 34.4	20 - 10
26	37.1 x 35.7	18 - 9
27	38.4 x 37.0	17 - 8.5
28	39.8 x 38.4	16 - 8
29	41.1 x 39.7	15 - 7.5
30	42.5 x 41.1	14 - 7
31		
32		
33		
34		
35		
36		
37		
38		
39		
40		
41		
42		
43		
44		
45		
46		
47		
48		
49		
50		

MTG. HGT.	ANGLE MOUNTED-30° A x B	'C' DIM.	INTENSITY D	E	F
10	22.8 x 13.8	6.07	78	95	39
11	24.8 x 15.2	6.64	64	79	32
12	27.0 x 16.5	7.23	54	66	27
13	29.0 x 17.8	7.80	46	56	23
14	31.2 x 19.2	8.84	40	48	20
15	33.2 x 20.4	8.96	35	42	17
16	35.3 x 21.8	9.54	30	37	15
17	37.3 x 23.1	10.1	27	33	13
18	39.4 x 24.4	10.7	24	29	12
19	41.5 x 25.8	11.3	22	26	11
20	43.6 x 27.1	11.8	19	24	9.5
21	45.6 x 28.4	12.4	18	21	9
22	47.7 x 29.8	13.0	16	19	8
23	49.8 x 31.1	13.6	15	18	7.5
24	51.9 x 32.4	14.2	13	16	6.5
25	53.9 x 33.8	14.7	12	15	6
26	56.0 x 35.0	15.3	11	14	5.5
27	58.0 x 36.4	15.9	10.6	13	5.3
28	60.2 x 37.7	16.5	9.8	12	4.9
29	62.3 x 39.0	17.0	9.2	11	4.6
30	64.4 x 40.4	17.6	8.6	10	4.3
31					
32					
33					
34					
35					

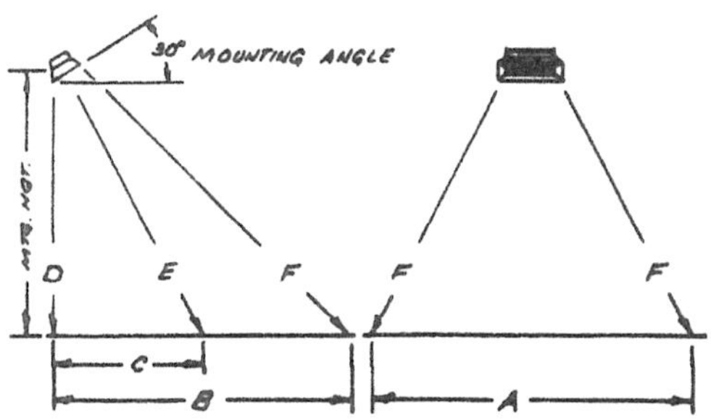

Table 12.4 Area Coverage for Mounted Heaters.

Solaronics

AREA COVERAGE

HEATER TYPE: 60 E BTU/HR INPUT: 50,000 GAS: NAT.

FOR HORIZONTAL MOUNTED HEATERS, A=LENGTH, B=WIDTH OF THE RECTANGLE COVERED

FOR ANGLE MOUNTED HEATERS, SEE THE DIAGRAM BELOW

MTG. HGT.	HORIZONTAL A x B	INTENSITY ℄ - EDGE
10	14.4 x 6.7	340 - 170
11	15.7 x 7.3	282 - 141
12	17.0 x 7.9	236 - 118
13	18.3 x 8.5	202 - 101
14	19.6 x 9.1	174 - 87
15	20.9 x 9.8	151 - 76
16	22.2 x 10.4	133 - 67
17	23.5 x 11.0	118 - 59
18	24.8 x 11.6	105 - 53
19	26.0 x 12.2	94 - 47
20	27.4 x 12.8	85 - 43
21	28.7 x 13.4	77 - 39
22	30.0 x 14.0	70 - 35
23	31.2 x 14.6	64 - 32
24	32.6 x 15.2	59 - 30
25	33.8 x 15.8	54 - 27
26	35.1 x 16.5	50 - 25
27	36.4 x 17.1	47 - 24
28	37.7 x 17.7	43 - 22
29	39.0 x 18.3	40 - 20
30	40.3 x 18.9	38 - 19
31	41.6 x 19.5	35 - 18
32	43.0 x 20.2	33 - 17
33	44.2 x 20.8	31 - 16
34	45.5 x 21.4	29 - 15
35	46.8 x 22.0	28 - 14
36	48.0 x 22.6	26 - 13
37	49.4 x 23.2	25 - 13
38	50.6 x 23.8	23 - 12
39	52.0 x 24.4	23 - 11
40	53.3 x 25.0	21 - 10
41	54.6 x 25.6	20 - 10
42	56.0 x 26.2	19 - 9
43	57.1 x 26.8	18 - 9
44	58.5 x 27.4	17.6 - 8.8
45	59.7 x 28.0	16.8 - 8.4
46	61.0 x 28.6	16 - 8
47	62.4 x 29.2	15.4 - 7.7
48	63.6 x 29.9	14.6 - 7.3
49	65.0 x 30.5	14 - 7
50	66.2 x 31.1	13.6 - 6.8

MTG. HGT.	ANGLE MOUNTED 30° A x B	'C' DIM.	INTENSITY D	E	F
10	23.6 x 13.8	6.07	140	255	70
11	25.8 x 15.2	6.64	116	210	58
12	27.9 x 16.5	7.23	97	178	49
13	30.0 x 17.8	7.80	83	152	42
14	32.3 x 19.2	8.84	71	130	36
15	34.4 x 20.4	8.96	62	114	31
16	36.5 x 21.8	9.54	55	99	28
17	38.6 x 23.1	10.1	48	88	24
18	40.7 x 24.4	10.7	43	78	22
19	43.3 x 25.8	11.3	39	71	20
20	45.1 x 27.1	11.8	35	64	18
21	47.2 x 28.4	12.4	32	57	16
22	49.4 x 29.8	13.0	29	53	15
23	51.5 x 31.1	13.6	27	48	14
24	53.7 x 32.4	14.2	24	44	12
25	55.8 x 33.8	14.7	22	41	11
26	58.0 x 35.0	15.3	21	38	10
27	60.0 x 36.4	15.9	19	35	9.5
28	62.3 x 37.7	16.5	18	32	9
29	64.5 x 39.0	17.0	17	30	8.5
30	66.8 x 40.4	17.6	16	28	8
31	69.0 x 41.8	18.2	14.6	26	7.3
32	71.0 x 43.1	18.8	13.8	25	6.9
33	73.0 x 44.4	19.3	12.8	23	6.4
34	75.2 x 45.7	19.9	12.2	22	6.1
35	77.6 x 47.0	20.5	11.4	21	5.7

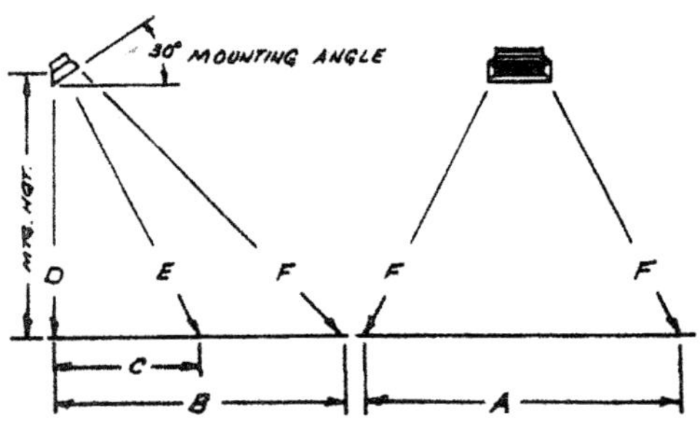

Table 12.5 Area Coverage for Mounted Heaters.

Solaronics

AREA COVERAGE

HEATER TYPE: 88 D-F BTU/HR INPUT: 75,000 GAS: NAT.

FOR HORIZONTAL MOUNTED HEATERS, A=LENGTH, B=WIDTH OF THE RECTANGLE COVERED

FOR ANGLE MOUNTED HEATERS, SEE THE DIAGRAM BELOW

MTG. HGT.	HORIZONTAL A x B	INTENSITY ℄ - EDGE
10	16.3 x 14.1	181 - 90
11	17.7 x 15.4	148 - 74
12	19.0 x 16.8	125 - 63
13	20.4 x 18.2	107 - 53
14	21.7 x 19.5	93 - 46
15	23.1 x 20.8	80 - 40
16	24.4 x 22.2	71 - 35
17	25.8 x 23.5	62 - 31
18	27.1 x 24.8	55 - 27
19	28.4 x 26.2	49 - 24
20	29.8 x 27.6	45 - 22
21	31.2 x 28.9	41 - 20
22	32.5 x 30.2	38 - 19
23	33.8 x 31.6	35 - 17
24	35.2 x 33.0	32 - 16
25	36.5 x 34.4	29 - 14
26	37.9 x 35.7	26 - 13
27	39.2 x 37.0	25 - 12
28	40.6 x 38.4	23 - 11
29	42.0 x 39.7	22 - 11
30	43.3 x 41.1	20 - 10
31		
32		
33		
34		
35		
36		
37		
38		
39		
40		
41		
42		
43		
44		
45		
46		
47		
48		
49		
50		

MTG. HGT.	ANGLE MOUNTED-30° A x B	'C' DIM.	INTENSITY D	E	F
10	23.6 x 13.8	6.07	113	138	56
11	25.7 x 15.2	6.64	93	115	46
12	27.8 x 16.5	7.23	78	96	39
13	29.8 x 17.8	7.80	67	81	33
14	32.0 x 19.2	8.48	58	70	29
15	34.0 x 20.4	8.96	51	61	25
16	36.1 x 21.8	9.54	43	54	21
17	38.1 x 23.1	10.1	39	48	19
18	40.2 x 24.4	10.7	35	42	17
19	42.3 x 25.8	11.3	32	38	16
20	44.4 x 27.1	11.8	28	35	14
21	46.4 x 28.4	12.4	26	30	13
22	48.5 x 29.8	13.0	23	28	12
23	50.6 x 31.1	13.6	22	26	11
24	52.7 x 32.4	14.2	19	23	9
25	54.7 x 33.8	14.7	17	22	8.5
26	56.8 x 35.0	15.3	16	20	8
27	58.8 x 36.4	15.9	15	19	7.5
28	61.0 x 37.7	16.5	14	17	7
29	63.1 x 39.0	17.0	13	16	6.5
30	65.2 x 40.4	17.6	12	15	6
31					
32					
33					
34					
35					

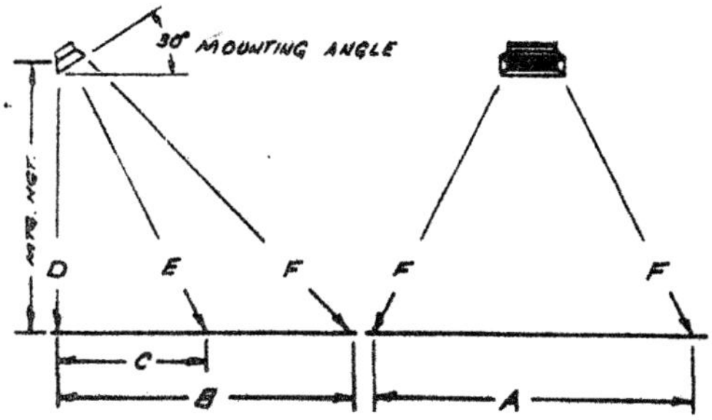

Table 12.6 Area Coverage for Mounted Heaters.

Solarnnics

AREA COVERAGE

HEATER TYPE: 88 E BTU/HR INPUT: 75,000 GAS: NAT.

FOR HORIZONTAL MOUNTED HEATERS, A=LENGTH, B=WIDTH OF THE RECTANGLE COVERED
FOR ANGLE MOUNTED HEATERS, SEE DIAGRAM BELOW

MTG. HGT.	HORIZONTAL A x B	INTENSITY ℄ - EDGE
10	15.8 x 6.7	462 - 231
11	17.1 x 7.3	384 - 192
12	18.4 x 7.9	321 - 160
13	19.7 x 8.5	275 - 137
14	21.0 x 9.1	237 - 118
15	22.3 x 99.8	206 - 103
16	23.6 x 10.4	181 - 90
17	24.9 x 11.0	161 - 80
18	26.2 x 11.6	143 - 71
19	27.4 x 12.2	128 - 64
20	28.8 x 12.8	116 - 58
21	30.1 x 13.4	105 - 52
22	31.4 x 14.0	95 - 47
23	32.6 x 14.6	87 - 43
24	34.0 x 15.2	80 - 40
25	35.2 x 15.8	74 - 37
26	36.5 x 16.5	68 - 34
27	37.8 x 17.1	64 - 32
28	39.1 x 17.7	58 - 29
29	40.4 x 18.3	54 - 27
30	41.7 x 18.9	52 - 26
31	43.0 x 19.5	48 - 24
32	44.4 x 20.2	45 - 22
33	45.6 x 20.8	42 - 21
34	46.9 x 21.4	39 - 20
35	48.2 x 22.0	38 - 19
36	49.4 x 22.6	35 - 18
37	50.8 x 23.2	34 - 17
38	52.0 x 23.8	32 - 16
39	53.4 x 24.4	31 - 15
40	54.7 x 25.0	29 - 14
41	56.0 x 25.6	27 - 13
42	57.4 x 26.2	26 - 13
43	58.5 x 26.8	25 - 12
44	59.9 x 27.4	24 - 12
45	61.1 x 28.0	23 - 11
46	62.4 x 28.6	22 - 11
47	63.8 x 29.2	21 - 10
48	65.0 x 29.9	20 - 10
49	66.4 x 30.5	19 - 9.5
50	67.6 x 31.1	18 - 9

MTG. HGT.	ANGLE MOUNTED-30° A x B	'C' DIM.	INTENSITY D	E	F
10	24.4 x 13.8	6.07	190	347	95
11	26.5 x 15.2	6.64	158	286	79
12	28.6 x 16.5	7.23	132	242	67
13	30.7 x 17.8	7.80	113	206	57
14	33.0 x 19.2	8.48	97	177	49
15	35.1 x 20.4	8.96	84	155	42
16	37.2 x 21.8	9.54	75	135	38
17	39.3 x 23.1	10.1	65	120	33
18	41.4 x 24.4	10.7	59	106	30
19	43.7 x 25.8	11.3	53	97	27
20	45.8 x 27.1	11.8	48	87	24
21	47.9 x 28.4	12.4	44	78	22
22	50.1 x 29.8	13.0	40	72	20
23	52.2 x 31.1	13.6	37	65	19
24	54.4 x 32.4	14.2	33	60	16
25	56.5 x 33.8	14.7	30	56	15
26	58.7 x 35.0	15.3	29	52	14
27	60.7 x 36.4	15.9	26	48	13
28	63.0 x 37.7	16.5	24	44	12
29	65.2 x 39.0	17.0	23	41	11.5
30	67.4 x 40.4	17.6	22	38	11
31	69.7 x 41.8	18.2	20	35	10
32	71.7 x 43.1	18.8	19	34	9.5
33	73.7 x 44.4	19.3	17.4	31	8.7
34	75.9 x 45.7	19.9	16.6	30	8.3
35	78.3 x 47.0	20.5	15.5	29	7.7

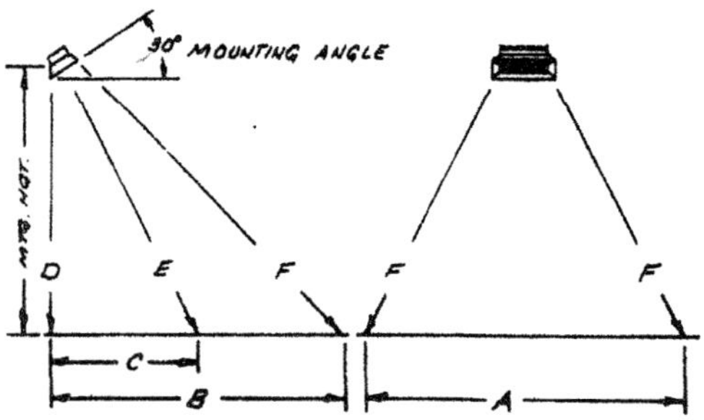

Table 12.7 Area Coverage for Mounted Heaters.

Solaronics

AREA COVERAGE

HEATER TYPE: 120 D-F BTU/HR INPUT: 100,000 GAS: NAT.

FOR HORIZONTAL MOUNTED HEATERS, A=LENGTH, B=WIDTH OF THE RECTANGLE COVERED

FOR ANGLE MOUNTED HEATERS= SEE THE DIAGRAM BELOW

MTG. HGT.	HORIZONTAL A x B	INTENSITY ℄ - EDGE
10	17.4 x 14.1	226 - 113
11	18.8 x 15.4	185 - 92
12	20.1 x 16.8	156 - 78
13	21.4 x 18.2	134 - 67
14	22.8 x 19.5	116 - 58
15	24.2 x 20.8	100 - 50
16	25.5 x 22.2	89 - 44
17	26.8 x 23.5	78 - 39
18	28.2 x 24.8	69 - 34
19	29.6 x 26.2	61 - 30
20	30.9 x 27.6	56 - 28
21	32.2 x 28.9	51 - 25
22	33.6 x 30.2	47 - 23
23	34.9 x 31.6	43 - 21
24	36.3 x 33.0	40 - 20
25	37.6 x 34.4	36 - 18
26	39.0 x 35.7	33 - 16
27	40.3 x 37.0	31 - 15
28	41.7 x 38.4	29 - 14
29	43.0 x 39.7	27 - 13
30	44.4 x 41.1	25 - 12
31		
32		
33		
34		
35		
36		
37		
38		
39		
40		
41		
42		
43		
44		
45		
46		
47		
48		
49		
50		

MTG. HGT.	ANGLE MOUNTED-30° A x B	'C' DIM.	INTENSITY D	E	F
10	24.8 x 13.8	6.07	141	172	71
11	26.8 x 15.2	6.64	116	143	58
12	29.0 x 16.5	7.23	98	120	49
13	31.0 x 17.8	7.80	83	101	42
14	33.2 x 19.2	8.48	72	87	36
15	35.2 x 20.4	8.96	63	76	31
16	37.3 x 21.8	9.54	54	67	27
17	39.3 x 23.1	10.1	49	60	24
18	41.4 x 24.4	10.7	43	53	22
19	43.5 x 25.8	11.3	40	47	20
20	45.6 x 27.1	11.8	34	44	17
21	47.6 x 28.4	12.4	33	38	16
22	49.7 x 29.8	13.0	29	34	14
23	51.8 x 31.1	13.6	27	33	13
24	53.9 x 32.4	14.2	24	29	12
25	55.9 x 33.8	14.7	22	27	11
26	58.0 x 35.0	15.3	20	25	10
27	60.0 x 36.4	15.9	19	24	9
28	62.2 x 37.7	16.5	18	22	9
29	64.3 x 39.0	17.0	17	20	8
30	66.4 x 40.4	17.6	16	18	8
31					
32					
33					
34					
35					

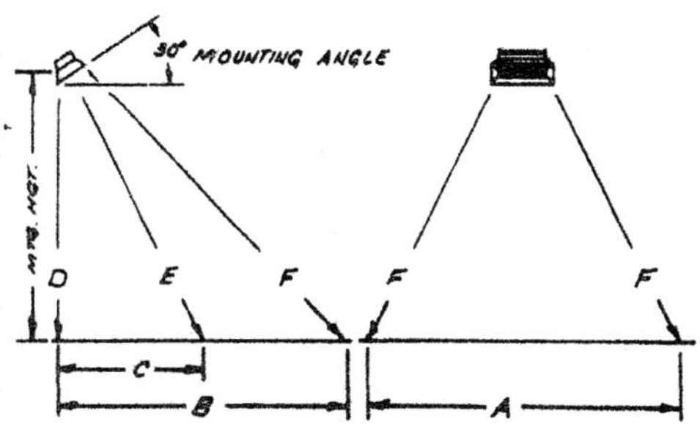

Table 12.8 Area Coverage for Mounted Heaters.

Solaronics
5-66

3 B
Page 9

AREA COVERAGE

HEATER TYPE: 120 E   BTU, HR INPUT: 100, 000   GAS: NAT.

FOR HORIZONTAL MOUNTED HEATERS, A=LENGTH, B=WIDTH OF THE RECTANGLE COVERED

FOR ANGLE MOUNTED HEATERS, SEE THE DIAGRAM BELOW

MTG. HGT.	HORIZONTAL A x B	INTENSITY ℄ - EDGE
10	17.0 x 6.7	575 - 287
11	18.3 x 7.3	475 - 237
12	19.6 x 7.9	400 - 200
13	20.9 x 8.5	340 - 170
14	22.2 x 9.8	293 - 146
15	23.5 x 9.8	255 - 127
16	24.8 x 10.4	224 - 112
17	26.1 x 11.0	198 - 99
18	27.4 x 11.6	178 - 89
19	28.6 x 12.2	159 - 79
20	30.0 x 12.8	144 - 72
21	31.3 x 13.4	130 - 65
22	32.6 x 14.0	119 - 60
23	33.8 x 14.6	108 - 54
24	35.2 x 15.2	100 - 50
25	36.4 x 15.8	92 - 46
26	37.7 x 16.5	85 - 43
27	39.0 x 17.1	79 - 40
28	40.3 x 17.7	73 - 336
29	41.6 x 18.3	68 - 34
30	42.9 x 18.9	64 - 32
31	44.2 x 19.5	60 - 30
32	45.6 x 20.2	56 - 28
33	46.8 x 20.8	53 - 26
34	48.1 x 21.4	50 - 25
35	49.4 x 22.0	47 - 24
36	50.6 x 22.6	44 - 22
37	52.0 x 23.2	42 - 21
38	53.2 x 23.8	40 - 20
39	54.6 x 24.4	38 - 19
40	55.9 x 25.0	36 - 18
41	57.2 x 25.6	34 - 17
42	58.6 x 26.2	33 - 16
43	59.7 x 26.8	31 - 15
44	61.1 x 27.4	30 - 15
45	62.3 x 28.0	28 - 14
46	63.3 x 28.6	27 - 13
47	65.0 x 29.2	26 - 13
48	66.2 x 29.9	25 - 12
49	67.6 x 30.5	24 - 12
50	68.8 x 31.1	23 - 11

MTG. HGT.	ANGLE MOUNTED-30°		INTENSITY		
	A x B	'C' DIM.	D	E	F
10	25.4 x 13.8	6.07	236	432	118
11	27.6 x 15.2	6.64	196	356	98
12	29.7 x 16.5	7.23	164	300	82
13	31.8 x 17.8	7.80	140	257	70
14	34.1 x 19.2	8.48	120	220	60
15	36.2 x 20.4	8.96	115	193	57
16	38.3 x 21.8	9.54	93	168	46
17	40.4 x 23.1	10.1	81	149	40
18	42.5 x 24.4	10.7	73	132	36
19	44.8 x 25.8	11.3	66	120	33
20	46.9 x 27.1	11.8	59	108	30
21	49.0 x 28.4	12.4	54	96	27
22	51.2 x 29.8	13.0	49	90	24
23	53.3 x 31.1	13.6	46	81	23
24	54.5 x 32.4	14.2	41	74	20
25	57.6 x 33.8	14.7	37	69	18
26	59.8 x 35.0	15.3	35	64	17
27	61.8 x 36.4	15.9	32	59	16
28	64.1 x 37.7	16.5	30	54	15
29	66.3 x 39.0	17.0	29	51	14
30	68.6 x 40.4	17.6	27	47	13
31	70.8 x 41.8	18.2	25	44	12
32	72.8 x 43.1	18.8	23	42	11
33	74.8 x 44.4	19.3	22	39	11
34	77.0 x 45.7	19.9	20	37	10
35	79.4 x 47.0	20.5	19	36	9

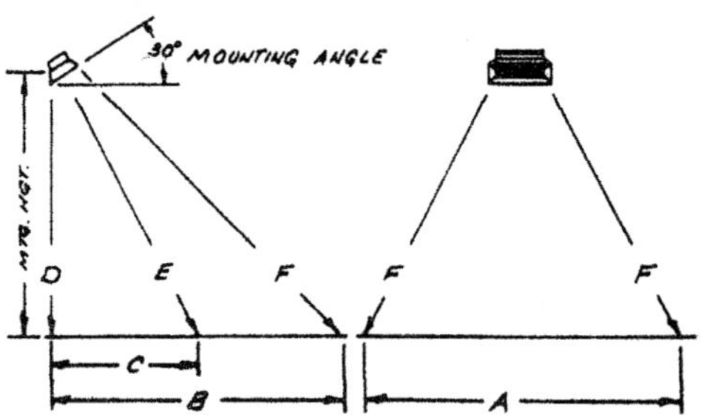

- Checkout areas of supermarkets or shopping centers where people and workers are exposed to open doors to the exits and entrances. Mostly for the checkout workers.
- Football and soccer stadiums that have reserved seating that may be sheltered from above but be affected by the elements or the weather.
- Enclosed loading or shipping docks that are literally outdoors, which have trucks or tractor-trailers that are backing into the enclosed space through large overhead or roll-up doors constantly opening and closing.
- Car washing facilities have rotating brushes that are wet that freeze-up in cold weather are great spot heating applications. Spot heating is also needed where the workers dry off the cars to keep them warm.

The above examples for spot or area heating are fairly obvious, but surely there are many more. Parking garages where you pay, automobile inspection stations where workers are exposed to the wind and the cold weather, baseball batting cages, ski lodges, patio heating, and construction sites are a few more that come to mind. Certainly, there are spot or area heating applications that you will find every day when you look around and use your imagination.

12.2 Portable IR Heaters

There is one other type of "spot" heating and that is using portable IR heaters. Portable IR heaters are sometimes referred to as construction heaters. The reason for that name is due to the fact that they are portable and can be used to cure concrete and provide heat for the various workers at a construction site. The IR units help keep the men on the job. They provide instant warmth and are ready in less than a minute. These portable IR heaters operate on liquid gas, propane at 11" water column pressure. They are useful for drying plaster, paint, or surfaces to be painted. Mainly they provide comfort heat for masons, plumbers, carpenters, electricians, plasterers, and everyone else working at the construction site.

The construction is not the only type of application for IR portable heater. The following are some of the more obvious types:

1. Railroad industry
    - For switch thawing
    - Warmth for track repair crews
    - For maintenance of equipment in the yards or open sheds

2. Trucking industry
    - Melt snow and ice on docks
    - Outdoor equipment repairs and service
    - Thaw frozen moisture in truck and trailer brake linings and drums
    - Warmth for workers doing repair and maintenance on equipment
3. Farms
    - Comfort heat while doing repair on tractors and other equipment
    - Thawing water pipes, fountains, etc.
4. Sporting activities
    - Warm football players on the sidelines
    - Skiers waiting for lifts, tow, or jump as well as spectators
    - Ice-fishermen
    - Golf practice tees, driving ranges
    - Duck blinds
5. Auto service
    - Roadside repairs
    - Grease rack
    - Wash rack
    - Brake drum thawing
6. Ready mix concrete plants
7. Outdoor shows
8. Welders
9. Municipalities
10. Utilities
11. Quarries

And the list goes on.

Conclusion

The primary purpose of this book was to illustrate just how effective gas-fired infra-red can be when compared with any other form of heating system. Even the oil industry has attempted, with the assistance of heating equipment manufacturers, to develop an oil-fired infra-red heater design. To date (at least to my knowledge), they have not been successful. Problems with oil creating soot in its combustion process will deny any possibility of an oil-fired infra-red heater. I personally feel that oil will go by the way of wood and coal except for areas where there is no natural gas or propane available. Even the electric utility industries are changing from oil to natural gas or nuclear fusion at their generating stations. The EPA also has much to say about how the byproducts of combustion affect the environment.

Overall energy costs will continue to increase as they have in the past. This will generally be a benefit to gas-fired infra-red heating. As has been pointed out in the text, HVAC engineers are going to look at new environmental regulations and code changes that will affect their thinking and designs. It may be necessary to combine various types of heating systems and equipment. Modular boilers and cooling units are concepts that may solve energy usage and efficiency problems as well as initial installed costs. Gas-fired infra-red equipment will always be able to justify its installation. There are many applications where modular boilers will be used for office heating with infra-red for plant operations. The modular boilers only come on as needed with boilers operating at full load for extended periods of time eliminating any short cycling that would reduce overall boiler efficiency. The same can be said with modular A/C systems. They would only come on as needed.

Another piece of equipment mentioned in the text is the energy recovery ventilators (ERVs), which can be combined with IR systems to even improve the IR operating costs and any ventilation requirements. Here exhaust air from the exhaust fans pass through the patented counter-flow ERV with the make-up air coming into the opposite end of the ERV. The final result is make-up air coming into the building within 2 °F of the inside temperature of

DOI: 10.1201/9788770046589-16

the space being ventilated. Below floor radiant heating with low temperature water from a boiler is also an available option. In any case, an infra-red heating design should always be a strong consideration when contemplating the best way for a client's requirement for heating of any kind. An economic analysis will always justify IR heating when infra-red is applicable.

There are many options available to the HVAC design engineer and all should be looked into. Gas-fired infra-red heating is not for every heating application but should definitely be considered for its obvious efficiency in building heating and spot/area heating applications.

Gas-fired infra-red heating is GREEN HEAT!

I hope you have found this book as an introduction to the "greenest" form of building heating–gas-fired infra-red. It is not meant to be the ultimate presentation on the design of infra-red systems, but hopefully be improved with any comments or suggestions from you, the reader.

Appendix: ASHRAE Design Temperature - Winter and Summer

SOLAR RADIATION DATA

Presently, there are 37 stations reporting solar radiation. Because of inconsistencies in the calibration of station measuring devices prior to 1975, the data were rehabilitated by the National Climatic Center.[11] The Canadian Climate Centre has prepared a merged solar and meteorological computer data base for 50 stations.[12]

WEATHER DATA RELATED TO EVAPORATIVE COOLING

ASHRAE RP 116 developed weather data for use in the design of evaporative cooling systems in the United States and Canada. Frequency of occurrence of median wet-bulb temperatures versus dry-bulb temperature data are available with design recommendations in Chapter 36 of the 1984 SYSTEMS VOLUME.

WEATHER DATA RELATED TO UNDERGROUND STRUCTURES

The National Bureau of Standards has accumulated and analyzed underground temperature data for a few stations in the contiguous United States.[16] NBS has measured earth temperatures beneath five different surfaces.[17]

TABLE 1 CLIMATIC CONDITIONS FOR THE UNITED STATES[a]

Col. 1	Col. 2		Col. 3		Col. 4	Winter,[d] F Col. 5		Summer,[e] F Col. 6			Col. 7	Col. 8		
State and Station	Latitude[b]		Longitude[b]		Elevation[c]	Design Dry-Bulb		Design Dry-Bulb and Mean Coincident Wet-Bulb			Mean Daily	Design Wet-Bulb		
	°	′	°	′	Ft	99%	97.5%	1%	2.5%	5%	Range	1%	2.5%	5%
ALABAMA														
Alexander City	33	0	86	0	660	18	22	96/77	93/76	91/76	21	79	78	78
Anniston AP	33	4	85	5	599	18	22	97/77	94/76	92/76	21	79	78	78
Auburn	32	4	85	3	730	18	22	96/77	93/76	91/76	21	79	78	78
Birmingham AP	33	3	86	5	610	17	21	96/74	94/75	92/74	21	78	77	76
Decatur	34	4	87	0	580	11	16	95/75	93/74	91/74	22	78	77	76
Dothan AP	31	2	85	2	321	23	27	94/76	92/76	91/76	20	80	79	78
Florence AP	34	5	87	4	528	17	21	97/74	94/74	92/74	22	78	77	76
Gadsden	34	0	86	0	570	16	20	96/75	94/75	92/74	22	78	77	76
Huntsville AP	34	4	86	4	619	11	16	95/75	93/74	91/74	23	78	77	76
Mobile AP	30	4	88	2	211	25	29	95/77	93/77	91/76	18	80	79	78
Mobile CO	30	4	88	1	119	25	29	95/77	93/77	91/76	16	80	79	78
Montgomery AP	32	2	86	2	195	22	25	96/76	95/76	93/76	21	79	79	78
Selma-Craig AFB	32	2	87	0	207	22	26	97/78	95/77	93/77	21	81	80	79
Talladega	33	3	86	1	565	18	22	97/77	94/76	92/76	21	79	78	78
Tuscaloosa AP	33	1	87	4	170r	20	23	98/75	96/76	94/76	22	79	78	77
ALASKA														
Anchorage AP	61	1	150	0	90	−23	−18	71/59	68/58	66/56	15	60	59	57
Barrow (S)	71	2	156	5	22	−45	−41	57/53	53/50	49/47	12	54	50	47
Fairbanks AP (S)	64	5	147	5	436	−51	−47	82/62	78/60	75/59	24	64	62	60
Juneau AP	58	2	134	4	17	−4	1	74/60	70/58	67/57	15	61	59	58
Kodiak	57	3	152	3	21	10	13	69/58	65/56	62/55	10	60	58	56
Nome AP	64	3	165	3	13	−31	−27	66/57	62/55	59/54	10	58	56	55
ARIZONA														
Douglas AP	31	3	109	3	4098	27	31	98/63	95/63	93/63	31	70	69	68
Flagstaff AP	35	1	111	4	6973	−2	4	84/55	82/55	80/54	31	61	60	59
Fort Huachuca AP (S)	31	3	110	2	4664	24	28	95/62	92/62	90/62	27	69	68	67
Kingman AP	35	2	114	0	3446	18	25	103/65	100/64	97/64	30	70	69	69
Nogales	31	2	111	0	3800	28	32	99/64	96/64	94/64	31	71	70	69
Phoenix AP(S)	33	3	112	0	1117	31	34	109/71	107/71	105/71	27	76	75	75
Prescott AP	34	4	112	3	5014	4	9	96/61	94/60	92/60	30	66	65	64
Tuscon AP (S)	32	1	111	0	2584	28	32	104/66	102/66	100/66	26	72	71	71
Winslow AP	35	0	110	4	4880	5	10	97/61	95/60	93/60	32	66	65	64
Yuma AP	32	4	114	4	199	36	39	111/72	109/72	107/71	27	79	78	77

[a] Table 1 was prepared by ASHRAE Technical Committee 4.2, Weather Data, from data compiled from official weather stations where hourly weather observations are made by trained observers, See also Ref. 3, 4, 5, 7, and 10.

[b] Latitude, for use in calculating solar loads, and longitude are given to the nearest 10minutes. For example, the latitude and longitude for Anniston, Alabama are given as 33 34 and 85 55 respectively, or 33° 40 and 85° 50.

[c] Elevations are ground elevations for each station. Temperature readings are generally made at an elevation of 5 ft above ground, except for locations marked r, indicating roof exposure of thermometer.

[d] Percentage of winter design data shows the percent of the 3-month period, December through February.

[e] Percentage of summer design data shows the percent of 4-month period, June through September.

TABLE 1 CLIMATIC CONDITIONS FOR THE UNITED STATES[a]

Col. 1	Col. 2		Col. 3		Col. 4	Col. 5 Winter,[d] F		Col. 6 Summer,[e] F			Col. 7	Col. 8		
State and Station	Latitude[b]		Longitude[b]		Elevation[c]	Design Dry-Bulb		Design Dry-Bulb and Mean Coincident Wet-Bulb			Mean Daily Range	Design Wet-Bulb		
	°	'	°	'	Ft	99%	97.5%	1%	2.5%	5%		1%	2.5%	5%
ARKANSAS														
Blytheville AFB	36	0	90	0	264	10	15	96/78	94/77	91/76	21	81	80	78
Camden	33	4	92	5	116	18	23	98/76	96/76	94/76	21	80	79	78
El Dorado AP	33	1	92	5	252	18	23	98/76	96/76	94/76	21	80	79	78
Fayetteville AP	36	0	94	1	1253	7	12	97/72	94/73	92/73	23	77	76	75
Fort Smith AP	35	2	94	2	449	12	17	101/75	98/76	95/76	24	80	79	78
Hot Springs	34	3	93	1	535	17	23	101/77	97/77	94/77	22	80	79	78
Jonesboro	35	5	90	4	345	10	15	96/78	94/77	91/76	21	81	80	78
Little Rock AP (S)	34	4	92	1	257	15	20	99/76	96/77	94/77	22	80	79	78
Pine Bluff AP	34	1	92	0	204	16	22	100/78	97/77	95/78	22	81	80	80
Texarkana AP	33	3	94	0	361	18	23	98/76	96/77	93/76	21	80	79	78
CALIFORNIA														
Bakersfield AP	35	2	119	0	495	30	32	104/70	101/69	98/68	32	73	71	70
Barstow AP	34	5	116	5	2142	26	29	106/68	104/68	102/67	37	73	71	70
Blythe AP	33	4	114	3	390	30	33	112/71	110/71	108/70	28	75	75	74
Burbank AP	34	1	118	2	699	37	39	95/68	91/68	88/67	25	71	70	69
Chico	39	5	121	5	205	28	30	103/69	101/68	98/67	36	71	70	68
Concord	38	0	122	0	195	24	27	100/69	97/68	94/67	32	71	70	68
Covina	34	0	117	5	575	32	35	98/69	95/68	92/67	31	73	71	70
Crescent City AP	41	5	124	0	50	31	33	68/60	65/59	63/58	18	62	60	59
Downey	34	0	118	1	116	37	40	93/70	89/70	86/69	22	72	71	70
El Cajon	32	4	117	0	525	42	44	83/69	80/69	78/68	30	71	70	68
El Centro AP (S)	32	5	115	4	−30	35	38	112/74	110/74	108/74	34	81	80	78
Escondido	33	0	117	1	660	39	41	89/68	85/68	82/68	30	71	70	69
Eureka/ Arcata AP	41	0	124	1	217	31	33	68/60	65/59	63/58	11	62	60	59
Fairfield- Travis AFB	38	2	122	0	72	29	32	99/68	95/67	91/66	34	70	68	67
Fresno AP (S)	36	5	119	4	326	28	30	102/70	100/69	97/68	34	72	71	70
Hamilton AFB	38	0	122	3	3	30	32	89/68	84/66	80/65	28	72	69	67
Laguna Beach	33	3	117	5	35	41	43	83/68	80/68	77/67	18	70	69	68
Livermore	37	4	122	0	545	24	27	100/69	97/68	93/67	24	71	70	68
Lompoc, Vandenburg AFB	34	4	120	3	552	35	38	75/61	70/61	67/60	20	63	61	60
Long Beach AP	33	5	118	1	34	41	43	83/68	80/68	77/67	22	70	69	68
Los Angeles AP (S)	34	0	118	2	99	41	43	83/68	80/68	77/67	15	70	69	68
Los Angeles CO (S)	34	0	118	1	312	37	40	93/70	89/70	86/69	20	72	71	70
Merced-Castle AFB	37	2	120	3	178	29	31	102/70	99/69	96/68	36	72	71	70
Modesto	37	4	121	0	91	28	30	101/69	98/68	95/67	36	71	70	69
Monterey	36	4	121	5	38	35	38	75/63	71/61	68/61	20	64	62	61
Napa	38	2	122	2	16	30	32	100/69	96/68	92/67	30	71	69	68
Needles AP	34	5	114	4	913	30	33	112/71	110/71	108/70	27	75	75	74
Oakland AP	37	4	122	1	3	34	36	85/64	80/63	75/62	19	66	64	63
Oceanside	33	1	117	2	30	41	43	83/68	80/68	77/67	13	70	69	68
Ontario	34	0	117	36	995	31	33	102/70	99/69	96/67	36	74	72	71
Obhard	34	1	119	1	43	34	36	83/66	80/64	77/63	19	70	68	67
Palmdale AP	34	4	118	1	2517	18	22	103/65	101/65	98/64	35	69	67	66
Palm Springs	33	5	116	4	411	33	35	112/71	110/70	108/70	35	76	74	73
Pasadena	34	1	118	1	864	32	35	98/69	95/68	92/67	29	73	71	70
Petaluma	38	1	122	4	27	26	29	94/68	90/66	87/65	31	72	70	68
Pomona CO	34	0	117	5	871	28	30	102/70	99/69	95/68	36	74	72	71
Redding AP	40	3	122	1	495	29	31	105/68	102/67	100/66	32	71	69	68
Redlands	34	0	117	1	1318	31	33	102/70	99/69	96/68	33	74	72	71
Richmond	38	0	122	2	55	34	36	85/64	80/63	75/62	17	66	64	63
Riverside- March AFB (S)	33	5	117	2	1511	29	32	100/68	98/68	95/67	37	72	71	70

TABLE 1 CLIMATIC CONDITIONS FOR THE UNITED STATES[a]

Col. 1	Col. 2		Col. 3		Col. 4	Col. 5 Winter,[d] F		Col. 6 Summer,[e] F			Col. 7	Col. 8		
State and Station	Latitude[b] °	'	Longitude[c] °	'	Elevation[c] Ft	Design Dry-Bulb 99%	97.5%	Design Dry-Bulb and Mean Coincident Wet-Bulb 1%	2.5%	5%	Mean Daily Range	Design Wet-Bulb 1%	2.5%	5%
Sacramento AP	38	3	121	3	17	30	32	101/70		94/69	36	72	71	98/70
Salinas AP	36	4	121	4	74	30	32	74/61	70/60	67/59	24	62	61	59
San Bernardino, Norton AFB	34	1	117	1	1125	31	33	102/70	99/69	96/68	38	74	72	71
San Diego AP	32	4	117	1	19	42	44	83/69	80/69	78/68	12	71	70	68
San Fernando	34	1	118	3	977	37	39	95/68	91/68	88/67	38	71	70	69
San Francisco AP	37	4	122	2	8	35	38	82/64	77/63	73/62	20	65	64	62
San Francisco CO	37	5	122	3	52	38	40	74/63	71/62	69/61	14	64	62	61
San Jose AP	37	2	122	0	70r	34	36	85/66	81/65	77/64	26	68	67	65
San Luis Obispo	35	2	120	4	315	33	35	92/69	88/70	84/69	26	73	71	70
Santa Ana AP	33	4	117	5	115r	37	39	89/69	85/68	82/68	28	71	70	69
Santa Barbara MAP	34	3	119	5	10	34	36	81/67	77/66	75/65	24	68	67	66
Santa Cruz	37	0	122	0	125	35	38	75/63	71/61	68/61	28	64	62	61
Santa Maria AP (S)	34	5	120	3	238	31	33	81/64	76/63	73/62	23	65	64	63
Santa Monica CO	34	0	118	3	57	41	43	83/68	80/68	77/67	16	70	69	68
Santa Paula	34	2	119	0	263	33	35	90/68	86/67	84/66	36	71	69	68
Santa Rosa	38	3	122	5	167	27	29	99/68	95/67	91/66	34	70	68	67
Stockton AP	37	5	121	2	28	28	30	100/69	97/68	94/67	37	71	70	68
Ukiah	39	1	122	4	620	27	29	99/69	95/68	91/67	40	70	68	67
Visalia	36	2	119	1	354	28	30	102/70	100/69	97/68	38	72	71	70
Yreka	41	4	122	4	2625	13	17	95/65	92/64	89/63	38	67	65	64
Yuba City	39	1	121	4	70	29	31	104/68	101/67	99/66	36	71	69	68
COLORADO														
Alamosa AP	37	3	105	5	7536	−21	−16	84/57	82/57	80/57	35	62	61	60
Boulder	40	0	105	2	5385	−2	8	93/59	91/59	89/59	27	64	63	62
Colorado Springs AP	38	5	104	4	6173	−3	2	91/58	88/57	86/57	30	63	62	61
Denver AP	39	5	104	5	5283	−5	1	93/59	91/59	89/59	28	64	63	62
Durango	37	1	107	5	6550	−1	4	89/59	87/59	85/59	30	64	63	62
Fort Collins	40	4	105	0	5001	−10	−4	93/59	91/59	89/59	28	64	63	62
Grand Junction AP (S)	39	1	108	3	4849	2	7	96/59	94/59	92/59	29	64	63	62
Greeley	40	3	104	4	4648	−11	−5	96/60	94/60	92/60	29	65	64	63
La Junta AP	38	0	103	3	4188	−3	3	100/68	98/68	95/67	31	72	70	69
Leadville	39	2	106	2	10177	−8	−4	84/52	81/51	78/50	30	56	55	54
Pueblo AP	38	2	104	2	4639	−7	0	97/61	95/61	92/61	31	67	66	65
Sterling	40	4	103	1	3939	−7	−2	95/62	93/62	90/62	30	67	66	65
Trinidad AP	37	2	104	2	5746	−2	3	93/61	91/61	89/61	32	66	65	64
CONNECTICUT														
Bridgeport AP	41	1	73	1	7	6	9	86/73	84/71	81/70	18	75	74	73
Hartford, Brainard Field	41	5	72	4	15	3	7	91/74	88/73	85/72	22	77	75	74
New Haven AP	41	2	73	0	6	3	7	88/75	84/73	82/72	17	76	75	74
New London	41	2	72	1	60	5	9	88/73	85/72	83/71	16	76	75	74
Norwalk	41	1	73	3	37	6	9	86/73	84/71	81/70	19	75	74	73
Norwich	41	3	72	0	20	3	7	89/75	86/73	83/72	18	76	75	74
Waterbury	41	3	73	0	605	−4	2	88/73	85/71	82/70	21	75	74	72
Windsor Locks, Bradley Field (S)	42	0	72	4	169	0	4	91/74	88/72	85/71	22	76	75	73
DELAWARE														
Dover AFB	39	0	75	3	38	11	15	92/75	90/75	87/74	18	79	77	76
Wilmington AP	39	4	75	3	78	10	14	92/74	89/74	87/73	20	77	76	75
DISTRICT OF COLUMBIA														
Andrews AFB	38	5	76	5	279	10	14	92/75	90/74	87/73	18	78	76	75
Washington National AP	38	5	77	0	14	14	17	93/75	91/74	89/74	18	78	77	76
FLORIDA														
Belle Glade	26	4	80	4	16	41	44	92/76	91/76	89/76	16	79	78	78

Weather Data and Design Conditions

TABLE 1 CLIMATIC CONDITIONS FOR THE UNITED STATES[a]

Col. 1	Col. 2		Col. 3		Col. 4	Winter,[d] F Col. 5		Summer,[e] F Col. 6			Col. 7	Col. 8		
State and Station	Latitude[b]		Longitude[c]		Elevation[c]	Design Dry-Bulb		Design Dry-Bulb and Mean Coincident Wet-Bulb			Mean Daily Range	Design Wet-Bulb		
	°	′	°	′	Ft	99%	97.5%	1%	2.5%	5%		1%	2.5%	5%
Cape Kennedy AP	28	3	80	3	16	35	38	90/78	88/78	87/78	15	80	79	79
Daytona Beach AP	29	1	81	0	31	32	35	92/78	90/77	88/77	15	80	79	78
Fort Lauderdale	26	0	80	1	13	42	46	92/78	91/78	90/78	15	80	79	79
Fort Myers AP	26	4	81	5	13	41	44	93/78	92/78	91/77	18	80	79	79
Fort Pierce	27	3	80	2	10	38	42	91/78	90/78	89/78	15	80	79	79
Gainesville AP (S)	29	4	82	2	155	28	31	95/77	93/77	92/77	18	80	79	78
Jacksonville AP	30	3	81	4	24	29	32	96/77	94/77	92/76	19	79	79	78
Key West AP	24	3	81	5	6	55	57	90/78	90/78	89/78	9	80	79	79
Lakeland CO (S)	28	0	82	0	214	39	41	93/76	91/76	89/76	17	79	78	78
Miami AP (S)	25	5	80	2	7	44	47	91/77	90/77	89/77	15	79	79	78
Miami Beach CO	25	5	80	1	9	45	48	90/77	89/77	88/77	10	79	79	78
Ocala	29	1	82	1	86	31	34	95/77	93/77	92/76	18	80	79	78
Orlando AP	28	3	81	2	106r	35	38	94/76	93/76	91/76	17	79	78	78
Panama City, Tyndall AFB	30	0	85	4	22	29	33	92/78	90/77	89/77	14	81	80	79
Pensacola CO	30	3	87	1	13	25	29	94/77	93/77	91/77	14	80	79	79
St. Augustine	29	5	81	2	15	31	35	92/78	89/78	87/78	16	80	79	79
St. Petersburg	28	0	82	4	35	36	40	92/77	91/77	90/76	16	79	79	78
Sanford	28	5	81	2	14	35	38	94/76	93/76	91/76	17	79	78	78
Sarasota	27	2	82	3	30	39	42	93/77	92/77	90/76	17	79	79	78
Tallahassee AP (S)	30	2	84	2	58	27	30	94/77	92/76	90/76	19	79	78	78
Tampa AP (S)	28	0	82	3	19	36	40	92/77	91/77	90/76	17	79	79	78
West Palm Beach AP	26	4	80	1	15	41	45	92/78	91/78	90/78	16	80	79	79
GEORGIA														
Albany, Turner AFB	31	3	84	1	224	25	29	97/77	95/76	93/76	20	80	79	78
Americus	32	0	84	2	476	21	25	97/77	94/76	92/75	20	79	78	77
Athens	34	0	83	2	700	18	22	94/74	92/74	90/74	21	78	77	76
Atlanta AP (S)	33	4	84	3	1005	17	22	94/74	92/74	90/73	19	77	76	75
Augusta AP	33	2	82	0	143	20	23	97/77	95/76	93/76	19	80	79	78
Brunswick	31	1	81	3	14	29	32	92/78	89/78	87/78	18	80	79	79
Columbus, Lawson AFB	32	3	85	0	242	21	24	95/76	93/76	91/75	21	79	78	77
Dalton	34	5	85	0	720	17	22	94/76	93/76	91/76	22	79	78	77
Dublin	32	3	83	0	215	21	25	96/77	93/76	91/75	20	79	78	77
Gainesville	34	2	83	5	1254	16	21	93/74	91/74	89/73	21	77	76	75
Griffin (S)	33	1	84	2	980	18	22	93/76	90/75	88/74	21	78	77	76
La Grange	33	0	85	0	715	19	23	94/76	91/75	89/74	21	78	77	76
Macon AP	32	4	83	4	356	21	25	96/77	93/76	91/75	22	79	78	77
Marietta, Dobbins AFB	34	0	84	3	1016	17	21	94/74	92/74	90/74	21	78	77	76
Moultrie	31	1	83	4	340	27	30	97/77	95/77	92/76	20	80	79	78
Rome AP	34	2	85	1	637	17	22	94/76	93/76	91/76	23	79	78	77
Savannah-Travis AP	32	1	81	1	52	24	27	96/77	93/77	91/77	20	80	79	78
Valdosta-Moody AFB	31	0	83	1	239	28	31	96/77	94/77	92/76	20	80	79	78
Waycross	31	2	82	2	140	26	29	96/77	94/77	91/76	20	80	79	78
HAWAII														
Hilo AP (S)	19	4	155	1	31	61	62	84/73	83/72	82/72	15	75	74	74
Honolulu AP	21	2	158	0	7	62	63	87/73	86/73	85/72	12	76	75	74
Kaneohe Bay MCAS	21	2	157	5	18	65	66	85/75	84/74	83/74	12	76	76	75
Wahiawa	21	3	158	0	900	58	59	86/73	85/72	84/72	14	75	74	73
IDAHO														
Boise AP(S)	43	3	116	1	2842	3	10	96/65	94/64	91/64	31	68	66	65
Burley	42	3	113	5	4180	−3	2	99/62	95/61	92/66	35	64	63	61
Coeur d'Alene AP	47	5	116	5	2973	−8	−1	89/62	86/61	83/60	31	64	63	61
Idaho Falls AP	43	3	112	0	4730r	−11	−6	89/61	87/61	84/59	38	65	63	61
Lewiston AP	46	2	117	0	1413	−1	6	96/65	93/64	90/63	32	67	66	64
Moscow	46	4	117	0	2660	−7	0	90/63	87/62	84/61	32	65	64	62
Mountain Home AFB	43	0	115	5	2992	6	12	99/64	97/63	94/62	36	66	65	63
Pocatello AP	43	0	112	4	4444	−8	−1	94/61	91/60	89/59	35	64	63	61
Twin Falls AP (S)	42	3	114	3	4148	−3	2	99/62	95/61	92/60	34	64	63	61

CHAPTER 24 **1985 Fundamentals Handbook**

TABLE 1 CLIMATIC CONDITIONS FOR THE UNITED STATES[a]

Col. 1	Col. 2		Col. 3		Col. 4	Winter,[d] F Col. 5		Summer,[e] F Col. 6			Col. 7	Col. 8		
State and Station	Latitude[b]		Longitude[b]		Elevation[c]	Design Dry-Bulb		Design Dry-Bulb and Mean Coincident Wet-Bulb			Mean Daily	Design Wet-Bulb		
	°	'	°	'	Ft	99%	97.5%	1%	2.5%	5%	Range	1%	2.5%	5%
ILLINOIS														
Aurora	41	5	88	2	744	−6	−1	93/76	91/76	88/75	20	79	78	76
Belleville, Scott AFB	38	3	89	5	447	1	6	94/76	92/76	89/75	21	79	78	76
Bloomington	40	3	89	0	775	−6	−2	92/75	90/74	88/73	21	78	76	75
Carbondale	37	5	89	1	380	2	7	95/77	93/77	90/76	21	80	79	77
Champaign/Urbana	40	0	88	2	743	−3	2	95/75	92/74	90/73	21	78	77	75
Chicago, Midway AP	41	5	87	5	610	−5	0	94/74	91/73	88/72	20	77	75	74
Chicago, O'Hare AP	42	0	87	5	658	−8	−4	91/74	89/74	86/72	20	77	76	74
Chicago CO	41	5	87	4	594	−3	2	94/75	91/74	88/73	15	79	77	75
Danville	40	1	87	4	558	−4	1	93/75	90/74	88/73	21	78	77	75
Decatur	39	5	88	5	670	−3	2	94/75	91/74	88/73	21	78	77	75
Dixon	41	5	89	3	696	−7	−2	93/75	90/74	88/73	23	78	77	75
Elgin	42	0	88	2	820	−7	−2	91/75	88/74	86/73	21	78	77	75
Freeport	42	2	89	4	780	−9	−4	91/74	89/73	87/72	24	77	76	74
Galesburg	41	0	90	3	771	−7	−2	93/75	91/75	88/74	22	78	77	75
Greenville	39	0	89	2	563	−1	4	94/76	92/75	89/74	21	79	78	76
Joliet	41	3	88	1	588	−5	0	93/75	90/74	88/73	20	78	77	75
Kankakee	41	1	87	5	625	−4	1	93/75	90/74	88/73	21	78	77	75
La Salle/Peru	41	2	89	1	520	−7	−2	93/75	91/75	88/74	22	78	77	75
Macomb	40	3	90	4	702	−5	0	95/76	92/76	89/75	22	79	78	76
Moline AP	41	3	90	3	582	−9	−4	93/75	91/75	88/74	23	78	77	75
Mt Vernon	38	2	88	5	500	0	5	95/76	92/75	89/74	21	79	78	76
Peoria AP	40	4	89	4	652	−8	−4	91/75	89/74	87/73	22	78	76	75
Quincy AP	40	0	91	1	762	−2	3	96/76	93/76	90/76	22	80	78	77
Rantoul, Chanute AFB	40	2	88	1	740	−4	1	94/75	91/74	89/73	21	78	77	75
Rockford	42	1	89	0	724	−9	−4	91/74	89/73	87/72	24	77	76	74
Springfield AP	39	5	89	4	587	−3	2	94/75	92/74	89/74	21	79	77	76
Waukegan	42	2	87	5	680	−6	−3	92/76	89/74	87/73	21	78	76	75
INDIANA														
Anderson	40	0	85	4	847	0	6	95/76	92/75	89/74	22	79	78	76
Bedford	38	5	86	3	670	0	5	95/76	92/75	89/74	22	79	78	76
Bloomington	39	1	86	4	820	0	5	95/76	92/75	89/74	22	79	78	76
Columbus, Bakalar AFB	39	2	85	5	661	3	7	95/76	92/75	90/74	22	79	78	76
Crawfordsville	40	0	86	5	752	−2	3	94/75	91/74	88/73	22	79	77	76
Evansville AP	38	0	87	3	381	4	9	95/76	93/75	91/75	22	79	78	77
Fort Wayne AP	41	0	85	1	791	−4	1	92/73	89/72	87/72	24	77	75	74
Goshen AP	41	3	85	5	823	−3	1	91/73	89/73	86/72	23	77	75	74
Hobart	41	3	87	2	600	−4	2	91/73	88/73	85/72	21	77	75	74
Huntington	40	4	85	3	802	−4	1	92/73	89/72	87/72	23	77	75	74
Indianapolis AP (S)	39	4	86	2	793	−2	2	92/74	90/74	87/73	22	78	76	75
Jeffersonville	38	2	85	5	455	5	10	95/74	93/74	90/74	23	79	77	76
Kokomo	40	3	86	1	790	−4	0	91/74	90/73	88/73	22	77	75	74
Lafayette	40	2	86	5	600	−3	3	94/74	91/73	88/73	22	78	76	75
La Porte	41	3	86	4	810	−3	3	93/74	90/74	87/73	22	78	76	75
Marion	40	3	85	4	791	−4	0	91/74	90/73	88/73	23	77	75	74
Muncie	40	1	85	2	955	−3	2	92/74	90/73	87/73	22	78	76	75
Peru, Grissom AFB	40	4	86	1	804	−6	−1	90/74	88/73	86/73	22	77	75	74
Richmond AP	39	5	84	5	1138	−2	2	92/74	90/74	87/73	22	78	76	75
Shelbyville	39	3	85	5	765	−1	3	93/74	91/74	88/73	22	78	76	75
South Bend AP	41	4	86	2	773	−3	1	91/73	89/73	86/72	22	77	75	74
Terre Haute AP	39	3	87	2	601	−2	4	95/75	92/74	89/73	22	79	77	76
Valparaiso	41	2	87	0	801	−3	3	93/74	90/74	87/73	22	78	76	75
Vincennes	38	4	87	3	420	1	6	95/75	92/74	90/73	22	79	77	76
IOWA														
Ames (S)	42	0	93	4	1004	−11	−6	93/75	90/74	87/73	23	78	76	75
Burlington AP	40	5	91	1	694	−7	−3	94/74	91/75	88/73	22	78	77	75
Cedar Rapids AP	41	5	91	4	863	−10	−5	91/76	88/75	86/74	23	78	77	75
Clinton	41	5	90	1	595	−8	−3	92/75	90/75	87/74	23	78	77	75
Council Bluffs	41	2	95	5	1210	−8	−3	94/76	91/75	88/74	22	78	77	75

Weather Data and Design Conditions

TABLE 1 CLIMATIC CONDITIONS FOR THE UNITED STATES[a]

						Winter,[d] F		Summer,[e] F						
Col. 1	Col. 2		Col. 3		Col. 4	Col. 5		Col. 6			Col. 7	Col. 8		
State and Station	Latitude[b]		Longitude[c]		Elevation[c]	Design Dry-Bulb		Design Dry-Bulb and Mean Coincident Wet-Bulb			Mean Daily Range	Design Wet-Bulb		
	°	′	°	′	Ft	99%	97.5%	1%	2.5%	5%		1%	2.5%	5%
Des Moines AP	41	3	93	4	948r	−10	−5	94/75	91/74	88/73	23	78	77	75
Dubuque	42	2	90	4	1065	−12	−7	90/74	88/73	86/72	22	77	75	74
Fort Dodge	42	3	94	1	1111	−12	−7	91/74	88/74	86/72	23	77	75	74
Iowa City	41	4	91	3	645	−11	−6	92/76	89/76	87/74	22	80	78	76
Keokuk	40	2	91	2	526	−5	0	95/75	92/75	89/74	22	79	77	76
Marshalltown	42	0	92	5	898	−12	−7	92/76	90/75	88/74	23	78	77	75
Mason City AP	43	1	93	2	1194	−15	−11	90/74	88/74	85/72	24	77	75	74
Newton	41	4	93	0	946	−10	−5	94/75	91/74	88/73	23	78	77	75
Ottumwa AP	41	1	92	2	842	−8	−4	94/75	91/74	88/73	22	78	77	75
Sioux City AP	42	2	96	2	1095	−11	−7	95/74	92/74	89/73	24	78	77	75
Waterloo	42	3	92	2	868	−15	−10	91/76	89/75	86/74	23	78	77	75
KANSAS														
Atchison	39	3	95	1	945	−2	2	96/77	93/76	91/76	23	81	79	77
Chanute AP	34	4	95	3	977	3	7	100/74	97/74	94/74	23	78	77	76
Dodge City AP (S)	37	5	100	0	2594	0	5	100/69	97/69	95/69	25	74	73	71
El Dorado	37	5	96	5	1282	3	7	101/72	98/73	96/73	24	77	76	75
Emporia	38	2	96	1	1209	1	5	100/74	97/74	94/73	25	78	77	76
Garden City AP	38	0	101	0	2882	−1	4	99/69	96/69	94/69	28	74	73	71
Goodland AP	39	2	101	4	3645	−5	0	99/66	96/65	93/66	31	71	70	68
Great Bend	38	2	98	5	1940	0	4	101/73	98/73	95/73	28	78	76	75
Hutchinson AP	38	0	97	5	1524	4	8	102/72	99/72	97/72	28	77	75	74
Liberal	37	0	101	0	2838	2	7	99/68	96/68	94/68	28	73	72	71
Manhattan, Fort Riley (S)	39	0	96	5	1076	−1	3	99/75	95/75	92/74	24	78	77	76
Parsons	37	2	95	3	908	5	9	100/74	97/74	94/74	23	79	77	76
Russell AP	38	5	98	5	1864	0	4	101/73	98/73	95/73	29	78	76	75
Salina	38	5	97	4	1271	0	5	103/74	100/74	97/73	26	78	77	75
Topeka AP	39	0	95	4	877	0	4	99/75	96/75	93/74	24	79	78	76
Wichita AP	37	4	97	3	1321	3	7	101/72	98/73	96/73	23	77	76	75
KENTUCKY														
Ashland	38	3	82	4	551	5	10	94/76	91/74	89/73	22	78	77	75
Bowling Green AP	37	0	86	3	535	4	10	94/77	92/75	89/74	21	79	77	76
Corbin AP	37	0	84	1	1175	4	9	94/73	92/73	89/72	23	77	76	75
Covington AP	39	0	84	4	869	1	6	92/73	90/72	88/72	22	77	75	74
Hopkinsville, Ft. Campbell	36	4	87	3	540	4	10	94/77	92/75	89/74	21	79	77	76
Lexington AP (S)	38	0	84	4	979	3	8	93/73	91/73	88/72	22	77	76	75
Louisville AP	38	1	85	4	474	5	10	95/74	93/74	90/74	23	79	77	76
Madisonville	37	2	87	3	439	5	10	96/76	93/75	90/75	22	79	78	77
Owensboro	37	5	87	1	420	5	10	97/76	94/75	91/75	23	79	78	77
Paducah AP	37	0	88	4	398	7	12	98/76	95/75	92/75	20	79	78	77
LOUISIANA														
Alexandria AP	31	2	92	2	92	23	27	95/77	94/77	92/77	20	80	79	78
Baton Rouge AP	30	3	91	1	64	25	29	95/77	93/77	92/77	19	80	80	79
Bogalusa	30	5	89	5	103	24	28	95/77	93/77	92/77	19	80	80	79
Houma	29	3	90	4	13	31	35	95/78	93/78	92/77	15	81	80	79
Lafayette AP	30	1	92	0	38	26	30	95/78	94/78	92/78	18	81	80	79
Lake Charles AP (S)	30	1	93	1	14	27	31	95/77	93/77	92/77	17	80	79	79
Minden	32	4	93	2	250	20	25	99/77	96/76	94/76	20	79	79	78
Monroe AP	32	3	92	0	78	20	25	99/77	96/76	94/76	20	79	79	78
Natchitoches	31	5	93	0	120	22	26	97/77	95/77	93/77	20	80	79	78
New Orleans AP	30	0	90	2	3	29	33	93/78	92/78	90/77	16	81	80	79
Shreveport AP(S)	32	3	93	5	252	20	25	99/77	96/76	94/76	20	79	79	78
MAINE														
Augusta AP	44	2	69	5	350	−7	−3	88/73	85/70	82/68	22	74	72	70
Bangor, Dow AFB	44	5	68	5	162	−11	−6	86/70	83/68	80/67	22	73	71	69
Caribou AP (S)	46	5	68	0	624	−18	−13	84/69	81/67	78/66	21	71	69	67
Lewiston	44	0	70	1	182	−7	−2	88/73	85/70	82/68	22	74	72	70
Millinocket AP	45	4	68	4	405	−13	−9	87/69	83/68	80/66	22	72	70	68
Portland (S)	43	4	70	2	61	−6	−1	87/72	84/71	81/69	22	74	72	70
Waterville	44	3	69	4	89	−8	−4	87/72	84/69	81/68	22	74	72	70

CHAPTER 24 **1985 Fundamentals Handbook**

TABLE 1 CLIMATIC CONDITIONS FOR THE UNITED STATES[a]

Col. 1	Col. 2		Col. 3		Col. 4	Winter,[d] F Col. 5		Summer,[e] F Col. 6			Col. 7	Col. 8		
State and Station	Latitude[b]		Longitude[c]		Elevation[c]	Design Dry-Bulb		Design Dry-Bulb and Mean Coincident Wet-Bulb			Mean Daily Range	Design Wet-Bulb		
	°	′	°	′	Ft	99%	97.5%	1%	2.5%	5%		1%	2.5%	5%
MARYLAND														
Baltimore AP	39	1	76	4	146	10	13	94/75	91/75	89/74	21	78	77	76
Baltimore CO	39	2	76	3	14	14	17	92/77	89/76	87/75	17	80	78	76
Cumberland	39	4	78	5	945	6	10	92/75	89/74	87/74	22	77	76	75
Frederick AP	39	3	77	3	294	8	12	94/76	91/75	88/74	22	78	77	76
Hagerstown	39	4	77	4	660	8	12	94/75	91/74	89/74	22	77	76	75
Salisbury (S)	38	2	75	3	52	12	16	93/75	91/75	88/74	18	79	77	76
MASSACHUSETTS														
Boston AP (S)	42	2	71	0	15	6	9	91/73	88/71	85/70	16	75	74	72
Clinton	42	2	71	4	398	−2	2	90/72	87/71	84/69	17	75	73	72
Fall River	41	4	71	1	190	5	9	87/72	84/71	81/69	18	74	73	72
Framingham	42	2	71	3	170	3	6	89/72	86/71	83/69	17	74	73	71
Gloucester	42	3	70	4	10	2	5	89/73	86/71	83/70	15	75	74	72
Greenfield	42	3	72	4	205	−7	−2	88/72	85/71	82/69	23	74	73	71
Lawrence	42	4	71	1	57	−6	0	90/73	87/72	84/70	22	76	74	73
Lowell	42	3	71	2	90	−4	1	91/73	88/72	85/70	21	76	74	73
New Bedford	41	4	71	0	70	5	9	85/72	82/71	80/69	19	74	73	72
Pittsfield AP	42	3	73	2	1170	−8	−3	87/71	84/70	81/68	23	73	72	70
Springfield, Westover AFB	42	1	72	3	247	−5	0	90/72	87/71	84/69	19	75	73	72
Taunton	41	5	71	1	20	5	9	89/73	86/72	83/70	18	75	74	73
Worcester AP	42	2	71	5	986	0	4	87/71	84/70	81/68	18	73	72	70
MICHIGAN														
Adrian	41	5	84	0	754	−1	3	91/73	88/72	85/71	23	76	75	73
Alpena AP	45	0	83	3	689	−11	−6	89/70	85/70	83/69	27	73	72	70
Battle Creek AP	42	2	85	2	939	1	5	92/74	88/72	85/70	23	76	74	73
Benton Harbor AP	42	1	86	3	649	1	5	91/72	88/72	85/70	20	75	74	72
Detroit	42	2	83	0	633	3	6	91/73	88/72	86/71	20	76	74	73
Escanaba	45	4	87	0	594	−11	−7	87/70	83/69	80/68	17	73	71	69
Flint AP	42	0	83	4	766	−4	1	90/73	87/72	85/71	25	76	74	72
Grand Rapids AP	42	5	85	3	681	1	5	91/72	88/72	85/70	24	75	74	72
Holland	42	5	86	1	612	2	6	88/72	86/71	83/70	22	75	73	72
Jackson AP	42	2	84	2	1003	1	5	92/74	88/72	85/70	23	76	74	73
Kalamazoo	42	1	85	3	930	1	5	92/74	88/72	85/70	23	76	74	73
Lansing AP	42	5	84	4	852	−3	1	90/73	87/72	84/70	24	75	74	72
Marquette CO	46	3	87	3	677	−12	−8	84/70	81/69	77/66	18	72	70	68
Mt Pleasant	43	4	84	5	796	0	4	91/73	87/72	84/71	24	76	74	72
Muskegon AP	43	1	86	1	627	2	6	86/72	84/70	82/70	21	75	73	72
Pontiac	42	4	83	2	974	0	4	90/73	87/72	85/71	21	76	74	73
Port Huron	43	0	82	3	586	0	4	90/73	87/72	83/71	21	76	74	73
Saginaw AP	43	3	84	1	662	0	4	91/73	87/72	84/71	23	76	74	72
Sault Ste. Marie AP (S)	46	3	84	2	721	−12	−8	84/70	81/69	77/66	23	72	70	68
Traverse City AP	44	4	85	4	618	−3	1	89/72	86/71	83/69	22	75	73	71
Ypsilanti	42	1	83	3	777	1	5	92/72	89/71	86/70	22	75	74	72
MINNESOTA														
Albert Lea	43	4	93	2	1235	−17	−12	90/74	87/72	84/71	24	77	75	73
Alexandria AP	45	5	95	2	1421	−22	−16	91/72	88/72	85/70	24	76	74	72
Bemidji AP	47	3	95	0	1392	−31	−26	88/69	85/69	81/67	24	73	71	69
Brainerd	46	2	94	2	1214	−20	−16	90/73	87/71	84/69	24	75	73	71
Duluth AP	46	5	92	1	1426	−21	−16	85/70	82/68	79/66	22	72	70	68
Fairbault	44	2	93	2	1190	−17	−12	91/74	88/72	85/71	24	77	75	73
Fergus Falls	46	1	96	0	1210	−21	−17	91/72	88/72	85/70	24	76	74	72
International Falls AP	48	3	93	2	1179	−29	−25	85/68	83/68	80/66	26	71	70	68
Mankato	44	1	94	0	785	−17	−12	91/72	88/72	85/70	24	77	75	73
Minneapolis/ St Paul AP	44	5	93	1	822	−16	−12	92/75	89/73	86/71	22	77	75	73
Rochester AP	44	0	92	3	1297	−17	−12	90/74	87/72	84/71	24	77	75	73
St Cloud AP (S)	45	4	94	1	1034	−15	−11	91/74	88/72	85/70	24	76	74	72

Weather Data and Design Conditions

TABLE 1 CLIMATIC CONDITIONS FOR THE UNITED STATES[a]

						Winter,[d] F		Summer,[e] F						
Col. 1	Col. 2		Col. 3		Col. 4	Col. 5		Col. 6			Col. 7	Col. 8		
State and Station	Latitude[b]		Longitude[b]		Elevation[c]	Design Dry-Bulb		Design Dry-Bulb and Mean Coincident Wet-Bulb			Mean Daily	Design Wet-Bulb		
	°	′	°	′	Ft	99%	97.5%	1%	2.5%	5%	Range	1%	2.5%	5%
Virginia	47	3	92	3	1435	−25	−21	85/69	83/68	80/66	23	71	70	68
Willmar	45	1	95	0	1133	−15	−11	91/74	88/72	85/71	24	76	74	72
Winona	44	1	91	4	652	−14	−10	91/75	88/73	85/72	24	77	75	74
MISSISSIPPI														
Biloxi, Keesler AFB	30	2	89	0	25	28	31	94/79	92/79	90/78	16	82	81	80
Clarksdale	34	1	90	3	178	14	19	96/77	94/77	92/76	21	80	79	78
Columbus AFB	33	4	88	3	224	15	20	95/77	93/77	91/76	22	80	79	78
Greenville AFB	33	3	91	1	139	15	20	95/77	93/77	91/76	21	80	79	78
Greenwood	33	3	90	1	128	15	20	95/77	93/77	91/76	21	80	79	78
Hattiesburg	31	2	89	2	200	24	27	96/78	94/77	92/77	21	81	80	79
Jackson AP	32	2	90	1	330	21	25	97/76	95/76	93/76	21	79	78	78
Laurel	31	4	89	1	264	24	27	96/78	94/77	92/77	21	81	80	79
McComb AP	31	2	90	3	458	21	26	96/77	94/76	92/76	18	80	79	78
Meridian AP	32	2	88	5	294	19	23	97/77	95/76	93/76	22	80	79	78
Natchez	31	4	91	3	168	23	27	96/78	94/78	92/77	21	81	80	79
Tupelo	34	2	88	4	289	14	19	96/77	94/77	92/76	22	80	79	78
Vicksburg CO	32	2	91	0	234	22	26	97/78	95/78	93/77	21	81	80	79
MISSOURI														
Cape Girardeau	37	1	89	3	330	8	13	98/76	95/75	92/75	21	79	78	77
Columbia AP (S)	39	0	92	2	778	−1	4	97/74	94/74	91/73	22	78	77	76
Farmington AP	37	5	90	3	928	3	8	96/76	93/75	90/74	22	78	77	75
Hannibal	39	4	91	2	489	−2	3	96/76	93/76	90/76	22	80	78	77
Jefferson City	38	4	92	1	640	2	7	98/75	95/74	92/74	23	78	77	76
Joplin AP	37	1	94	3	982	6	10	100/73	97/73	94/73	24	78	77	76
Kansas City AP	39	1	94	4	742	2	6	99/75	96/74	93/74	20	78	77	76
Kirksville AP	40	1	92	4	966	−5	0	96/74	93/74	90/73	24	78	77	76
Mexico	39	1	92	0	775	−1	4	97/74	94/74	91/73	22	78	77	76
Moberly	39	3	92	3	850	−2	3	97/74	94/74	91/73	23	78	77	76
Poplar Bluff	36	5	90	3	322	11	16	98/78	95/76	92/76	22	81	79	78
Rolla	38	0	91	5	1202	3	9	94/77	91/75	89/74	22	78	77	76
St Joseph AP	39	5	95	0	809	−3	2	96/77	93/76	91/76	23	81	79	77
St Louis AP	38	5	90	2	535	2	6	97/75	94/75	91/74	21	78	77	76
St Louis CO	38	4	90	2	465	3	8	98/75	94/75	91/74	18	78	77	76
Sedalia, Whiteman AFB	38	4	93	3	838	−1	4	95/76	92/76	90/75	22	79	78	76
Sikeston	36	5	89	3	318	9	15	98/77	95/76	92/75	21	80	78	77
Springfield AP	37	1	93	2	1265	3	9	96/73	93/74	91/74	23	78	77	75
MONTANA														
Billings AP	45	5	108	3	3567	−15	−10	94/64	91/64	88/63	31	67	66	64
Bozeman	45	5	111	0	4856	−20	−14	90/61	87/60	84/59	32	63	62	60
Butte AP	46	0	112	3	5526r	−24	−17	86/58	83/56	80/56	35	60	58	57
Cut Bank AP	48	4	112	2	3838r	−25	−20	88/61	85/61	82/60	35	64	62	61
Glasgow AP (S)	48	1	106	4	2277	−22	−18	92/64	89/63	85/62	29	68	66	64
Glendive	47	1	104	4	2076	−18	−13	95/66	92/64	89/62	29	69	67	65
Great Falls AP (S)	47	3	111	2	3664r	−21	−15	91/60	88/60	85/59	28	64	62	60
Havre	48	3	109	4	2488	−18	−11	94/65	90/64	87/63	33	68	66	65
Helena AP	46	4	112	0	3893	−21	−16	91/60	88/60	85/59	32	64	62	61
Kalispell AP	48	2	114	2	2965	−14	−7	91/62	87/61	84/60	34	65	63	62
Lewiston AP	47	0	109	3	4132	−22	−16	90/62	87/61	83/60	30	65	63	62
Livingston AP	45	4	110	3	4653	−20	−14	90/61	87/60	84/59	32	63	62	60
Miles City AP	46	3	105	5	2629	−20	−15	98/66	95/66	92/65	30	70	68	67
Missoula AP	46	5	114	1	3200	−13	−6	92/62	88/61	85/60	36	65	63	62
NEBRASKA														
Beatrice	40	2	96	5	1235	−5	−2	99/75	95/74	92/74	24	78	77	76
Chadron AP	42	5	103	0	3300	−8	−3	97/66	94/65	91/65	30	71	69	68
Columbus	41	3	97	2	1442	−6	−2	98/74	95/73	92/73	25	77	76	75
Fremont	41	3	96	3	1203	−6	−2	98/75	95/74	92/74	22	78	77	76

TABLE 1 CLIMATIC CONDITIONS FOR THE UNITED STATES[a]

Col. 1 State and Station	Col. 2 Latitude[b] °	'	Col. 3 Longitude[c] °	'	Col. 4 Elevation[c] Ft	Winter,[d] F Col. 5 Design Dry-Bulb 99%	97.5%	Summer,[e] F Col. 6 Design Dry-Bulb and Mean Coincident Wet-Bulb 1%	2.5%	5%	Col. 7 Mean Daily Range	Col. 8 Design Wet-Bulb 1%	2.5%	5%
Grand Island AP	41	0	98	2	1841	−8	−3	97/72	94/71	91/71	28	75	74	73
Hastings	40	4	98	3	1932	−7	−3	97/72	94/71	91/71	27	75	74	73
Kearney	40	4	99	1	2146	−9	−4	96/71	93/70	90/70	28	74	73	72
Lincoln CO (S)	40	5	96	5	1150	−5	−2	99/75	95/74	92/74	24	78	77	76
McCook	40	1	100	4	2565	−6	−2	98/69	95/69	91/69	28	74	72	71
Norfolk	42	0	97	3	1532	−8	−4	97/74	93/74	90/73	30	78	77	75
North Platte AP (S)	41	1	100	4	2779	−8	−4	97/69	94/69	90/69	28	74	72	71
Omaha AP	41	2	95	5	978	−8	−3	94/76	91/75	88/74	22	78	77	75
Scottsbluff AP	41	5	103	4	3950	−8	−3	95/65	92/65	90/64	31	70	68	67
Sidney AP	41	1	103	0	4292	−8	−3	95/65	92/65	90/64	31	70	68	67
NEVADA														
Carson City	39	1	119	5	4675	4	9	94/60	91/59	89/58	42	63	61	60
Elko AP	40	5	115	5	5075	−8	−2	94/59	92/59	90/58	42	63	62	60
Ely AP (S)	39	1	114	5	6257	−10	−4	89/57	87/56	85/55	39	60	59	58
Las Vegas AP (S)	36	1	115	1	2162	25	28	108/66	106/65	104/65	30	71	70	69
Lovelock AP	40	0	118	3	3900	8	12	98/63	96/63	93/62	42	66	65	64
Reno AP (S)	39	3	119	5	4404	5	10	95/61	92/60	90/59	45	64	62	61
Reno CO	39	3	119	5	4490	6	11	96/61	93/60	91/59	45	64	62	61
Tonopah AP	38	0	117	1	5426	5	10	94/60	92/59	90/58	40	64	62	61
Winnemucca AP	40	5	117	5	4299	−1	3	96/60	94/60	92/60	42	64	62	61
NEW HAMPSHIRE														
Berlin	44	3	71	1	1110	−14	−9	87/71	84/69	81/68	22	73	71	70
Claremont	43	2	72	2	420	−9	−4	89/72	86/70	83/69	24	74	73	71
Concord AP	43	1	71	3	339	−8	−3	90/72	87/70	84/69	26	74	73	71
Keene	43	0	72	2	490	−12	−7	90/72	87/70	83/69	24	74	73	71
Laconia	43	3	71	3	505	−10	−5	89/72	86/70	83/69	25	74	73	71
Manchester, Grenier AFB	43	0	71	3	253	−8	−3	91/72	88/71	85/70	24	75	74	72
Portsmouth, Pease AFB	43	1	70	5	127	−2	2	89/73	85/71	83/70	22	75	74	72
NEW JERSEY														
Atlantic City CO	39	3	74	3	11	10	13	92/74	89/74	86/72	18	78	77	75
Long Branch	40	2	74	0	20	10	13	93/74	90/73	87/72	18	78	77	75
Newark AP	40	4	74	1	11	10	14	94/74	91/73	88/72	20	77	76	75
New Brunswick	40	3	74	3	86	6	10	92/74	89/73	86/72	19	77	76	75
Paterson	40	5	74	1	100	6	10	94/74	91/73	88/72	21	77	76	75
Phillipsburg	40	4	75	1	180	1	6	92/73	89/72	86/71	21	76	75	74
Trenton CO	40	1	74	5	144	11	14	91/75	88/74	85/73	19	78	76	75
Vineland	39	3	75	0	95	8	11	91/75	89/74	86/73	19	78	76	75
NEW MEXICO														
Alamagordo, Holloman AFB	32	5	106	1	4070	14	19	98/64	96/64	94/64	30	69	68	67
Albuquerque AP (S)	35	0	106	4	5310	12	16	96/61	94/61	92/61	27	66	65	64
Artesia	32	5	104	2	3375	13	19	103/67	100/67	97/67	30	72	71	70
Carlsbad AP	32	2	104	2	3234	13	19	103/67	100/67	97/67	28	72	71	70
Clovis AP	34	3	103	1	4279	8	13	95/65	93/65	91/65	28	69	68	67
Farmington AP	36	5	108	1	5495	1	6	95/63	93/62	91/61	30	67	65	64
Gallup	35	3	108	5	6465	0	5	90/59	89/58	86/58	32	64	62	61
Grants	35	1	107	5	6520	−1	4	89/59	88/58	85/57	32	64	62	61
Hobbs AP	32	4	103	1	3664	13	18	101/66	99/66	97/66	29	71	70	69
Las Cruces	32	2	107	0	3900	15	20	99/64	96/64	94/64	30	69	68	67
Los Alamos	35	5	106	2	7410	5	9	89/60	87/60	85/60	32	62	61	60
Raton AP	36	5	104	3	6379	−4	1	91/60	89/60	87/60	34	65	64	63
Roswell, Walker AFB	33	2	104	3	3643	13	18	100/66	98/66	96/66	33	71	70	69
Santa Fe CO	35	4	106	0	7045	6	10	90/61	88/61	86/61	28	63	62	61
Silver City AP	32	4	108	2	5373	5	10	95/61	94/60	91/60	30	66	64	63
Socorro AP	34	0	106	5	4617	13	17	97/62	95/62	93/62	30	67	66	65
Tucumcari AP	35	1	103	4	4053	8	13	99/66	97/66	95/65	28	70	69	68
NEW YORK														
Albany AP (S)	42	5	73	5	277	−6	−1	91/73	88/72	85/70	23	75	74	72
Albany CO	42	5	73	5	19	−4	1	91/73	88/72	85/70	20	75	74	72
Auburn	43	0	76	3	715	−3	2	90/73	87/71	84/70	22	75	73	72
Batavia	43	0	78	1	900	1	5	90/72	87/71	84/70	22	75	73	72
Binghamton AP	42	1	76	0	1590	−2	1	86/71	83/69	81/68	20	73	72	70
Buffalo AP	43	0	78	4	705r	2	6	88/71	85/70	83/69	21	74	73	72
Cortland	42	4	76	1	1129	−5	0	88/71	85/71	82/70	23	74	73	71
Dunkirk	42	3	79	2	590	4	9	88/73	85/72	83/71	18	75	74	72

Weather Data and Design Conditions

TABLE 1 CLIMATIC CONDITIONS FOR THE UNITED STATES[a]

Col. 1	Col. 2		Col. 3		Col. 4	Winter,[d] F Col. 5		Summer,[e] F Col. 6			Col. 7	Col. 8		
State and Station	Latitude[b]		Longitude[c]		Elevation[c]	Design Dry-Bulb		Design Dry-Bulb and Mean Coincident Wet-Bulb			Mean Daily	Design Wet-Bulb		
	°	'	°	'	Ft	99%	97.5%	1%	2.5%	5%	Range	1%	2.5%	5%
Elmira AP	42	1	76	5	860	−4	1	89/71	86/71	83/70	24	74	73	71
Geneva (S)	42	5	77	0	590	−3	2	90/73	87/71	84/70	22	75	73	72
Glens Falls	42	2	73	4	321	−11	−5	88/72	85/71	82/69	23	74	73	71
Gloversville	43	1	74	2	790	−8	−2	89/72	86/71	83/69	23	75	74	72
Hornell	42	2	77	4	1325	−4	0	88/71	85/70	82/69	24	74	73	72
Ithaca (S)	42	3	76	3	950	−5	0	88/71	85/71	82/70	24	74	73	71
Jamestown	42	1	79	2	1390	−1	3	88/70	86/70	83/69	20	74	72	71
Kingston	42	0	74	0	279	−3	2	91/73	88/72	85/70	22	76	74	73
Lockport	43	1	78	4	520	4	7	89/74	86/72	84/71	21	76	74	73
Massena AP	45	0	75	0	202r	−13	−8	86/70	83/69	80/68	20	73	72	70
Newburg-Stewart AFB	41	3	74	1	460	−1	4	90/73	88/72	85/70	21	76	74	73
NYC-Central Park (S)	40	5	74	0	132	11	15	92/74	89/73	87/72	17	76	75	74
NYC-Kennedy AP	40	4	73	5	16	12	15	90/73	87/72	84/71	16	76	75	74
NYC-La Guardia AP	40	5	73	5	19	11	15	92/74	89/73	87/72	16	76	75	74
Niagra Falls AP	43	1	79	0	596	4	7	89/74	86/72	84/71	20	76	74	73
Olean	42	1	78	3	1420	−2	2	87/71	84/71	81/70	23	74	73	71
Oneonta	42	3	75	0	1150	−7	−4	86/71	83/69	80/68	24	73	72	70
Oswego CO	43	3	76	3	300	1	7	86/73	83/71	80/70	20	75	73	72
Plattsburg AFB	44	4	73	3	165	−13	−8	86/70	83/69	80/68	22	73	72	70
Poughkeepsie	41	4	73	5	103	0	6	92/74	89/74	86/72	21	77	75	74
Rochester AP	43	1	77	4	543	1	5	91/73	88/71	85/70	22	75	73	72
Rome-Griffiss AFB	43	1	75	3	515	−11	−5	88/71	85/70	83/69	22	75	73	71
Schenectady (S)	42	5	74	0	217	−4	1	90/73	87/72	84/70	22	75	74	72
Suffolk County AFB	40	5			57	7	10	86/72	83/71	80/70	16	76	74	73
Syracuse AP	43	1	76	1	424	−3	2	90/73	87/71	84/70	20	75	73	72
Utica	43	1	75	2	714	−12	−6	88/73	85/71	82/70	22	75	73	71
Watertown	44	0	76	0	497	−11	−6	86/73	83/71	81/70	20	75	73	72
NORTH CAROLINA														
Asheville AP	35	3	82	3	217r	10	14	89/73	87/72	85/71	21	75	74	72
Charlotte AP	35	0	81	0	735	18	22	95/74	93/74	91/74	20	77	76	76
Durham	36	0	78	5	406	16	20	94/75	92/75	90/75	20	78	77	76
Elizabeth City AP	36	2	76	1	10	12	19	93/78	91/77	89/76	18	80	78	78
Fayetteville, Pope AFB	35	1	79	0	95	17	20	95/76	92/76	90/75	20	79	78	77
Goldsboro, Seymour-Johnson AFB	35	2	78	0	88	18	21	94/77	91/76	89/75	18	79	78	77
Greensboro AP (S)	36	1	80	0	887	14	18	93/74	91/73	89/73	21	77	76	75
Greenville	35	4	77	2	25	18	21	93/77	91/76	89/75	19	79	78	77
Henderson	36	2	78	2	510	12	15	95/77	92/76	90/76	20	79	78	77
Hickory	35	4	81	2	1165	14	18	92/73	90/72	88/72	21	75	74	73
Jacksonville	34	5	77	3	24	20	24	92/78	90/78	88/77	18	80	79	78
Lumberton	34	4	79	0	132	18	21	95/76	92/76	90/75	20	79	78	77
New Bern AP	35	1	77	0	17	20	24	92/78	90/78	88/77	18	80	79	78
Raleigh/Durham AP (S)	35	5	78	5	433	16	20	94/75	92/75	90/75	20	78	77	76
Rocky Mount	36	0	77	5	81	18	21	94/77	91/76	89/75	19	79	78	77
Wilmington AP	34	2	78	0	30	23	26	93/79	91/78	89/77	18	81	80	79
Winston-Salem AP	36	1	80	1	967	16	20	94/74	91/73	89/73	20	76	75	74
NORTH DAKOTA														
Bismarck AP (S)	46	5	100	5	1647	−23	−19	95/68	91/68	88/67	27	73	71	70
Devil's Lake	48	1	98	5	1471	−25	−21	91/69	88/68	85/66	25	73	71	69
Dickinson AP	46	5	102	5	2595	−21	−17	94/68	90/66	87/65	25	71	69	68
Fargo AP	46	5	96	5	900	−22	−18	92/73	89/71	85/69	25	76	74	72
Grand Forks AP	48	0	97	2	832	−26	−22	91/70	87/70	84/68	25	74	72	70
Jamestown AP	47	0	98	4	1492	−22	−18	94/70	90/69	87/68	26	74	74	71
Minot AP	48	2	101	2	1713	−24	−20	92/68	89/67	86/65	25	72	70	68
Williston	48	1	103	4	1877	−25	−21	91/68	88/67	85/65	25	72	70	68
OHIO														
Akron-Canton AP	41	0	81	3	1210	1	6	89/72	86/71	84/70	21	75	73	72
Ashtabula	42	0	80	5	690	4	9	88/73	85/72	83/71	18	75	74	72
Athens	39	2	82	1	700	0	6	95/75	92/74	90/73	22	78	76	74
Bowling Green	41	3	83	4	675	−2	2	92/73	89/73	86/71	23	76	75	73

CHAPTER 24 **1985 Fundamentals Handbook**

TABLE 1 CLIMATIC CONDITIONS FOR THE UNITED STATES[a]

Col. 1	Col. 2		Col. 3		Col. 4	Col. 5 Winter,[d] F		Col. 6 Summer,[e] F			Col. 7	Col. 8		
State and Station	Latitude[b]		Longitude[c]		Elevation[c]	Design Dry-Bulb		Design Dry-Bulb and Mean Coincident Wet-Bulb			Mean Daily	Design Wet-Bulb		
	°	'	°	'	Ft	99%	97.5%	1%	2.5%	5%	Range	1%	2.5%	5%
Cambridge	40	0	81	4	800	1	7	93/75	90/74	87/73	23	78	76	75
Chillicothe	39	2	83	0	638	0	6	95/75	92/74	90/73	22	78	76	74
Cincinnati CO	39	1	84	4	761	1	6	92/73	90/72	88/72	21	77	75	74
Cleveland AP (S)	41	2	81	5	777r	1	5	91/73	88/72	86/71	22	76	74	73
Columbus AP (S)	40	0	82	5	812	0	5	92/73	90/73	87/72	24	77	75	74
Dayton AP	39	5	84	1	997	−1	4	91/73	89/72	86/71	20	76	75	73
Defiance	41	2	84	2	700	−1	4	94/74	91/73	88/72	24	77	76	74
Findlay AP	41	0	83	4	797	2	3	92/74	90/73	87/72	24	77	76	74
Fremont	41	2	83	1	600	−3	1	90/73	88/73	85/71	24	76	75	73
Hamilton	39	2	84	3	650	0	5	92/73	90/72	87/71	22	76	75	73
Lancaster	39	4	82	4	920	0	5	93/74	91/73	88/72	23	77	75	74
Lima	40	4	84	0	860	−1	4	94/74	91/73	88/72	24	77	76	74
Mansfield AP	40	5	82	3	1297	0	5	90/73	87/72	85/72	22	76	74	73
Marion	40	4	83	1	920	0	5	93/74	91/73	88/72	23	77	76	74
Middletown	39	3	84	3	635	0	5	92/73	90/72	87/71	22	76	75	73
Newark	40	1	82	3	825	−1	5	94/73	92/73	89/72	23	77	75	74
Norwalk	41	1	82	4	720	−3	1	90/73	88/73	85/71	22	76	75	73
Portsmouth	38	5	83	0	530	5	10	95/76	92/74	89/73	22	78	77	75
Sandusky CO	41	3	82	4	606	1	6	93/73	91/72	88/71	21	76	74	73
Springfield	40	0	83	5	1020	−1	3	91/74	89/73	87/72	21	77	76	74
Steubenville	40	2	80	4	992	1	5	89/72	86/71	84/70	22	74	73	72
Toledo AP	41	4	83	5	676r	−3	1	90/73	88/73	85/71	25	76	75	73
Warren	41	2	80	5	900	0	5	89/71	87/71	85/70	23	74	73	71
Wooster	40	5	82	0	1030	1	6	89/72	86/71	84/70	22	75	73	72
Youngstown AP	41	2	80	4	1178	−1	4	88/71	86/71	84/70	23	74	73	71
Zanesville AP	40	0	81	5	881	1	7	93/75	90/74	87/73	23	78	76	75
OKLAHOMA														
Ada	34	5	96	4	1015	10	14	100/74	97/74	95/74	23	77	76	75
Altus AFB	34	4	99	2	1390	11	16	102/73	100/73	98/73	25	77	76	75
Ardmore	34	2	97	1	880	13	17	100/74	98/74	95/74	23	77	77	76
Bartlesville	36	5	96	0	715	6	10	101/73	98/74	95/74	23	77	77	76
Chickasha	35	0	98	0	1085	10	14	101/74	98/74	95/74	24	78	77	76
Enid-Vance AFB	36	2	98	0	1287	9	13	103/74	100/74	97/74	24	79	77	76
Lawton AP	34	3	98	2	1108	12	16	101/74	99/74	96/74	24	78	77	76
Mc Alester	34	5	95	5	760	14	19	99/74	96/74	93/74	23	77	76	75
Muskogee AP	35	4	95	2	610	10	15	101/74	98/75	95/75	23	79	78	77
Norman	35	1	97	3	1109	9	13	99/74	96/74	94/74	24	77	76	75
Oklahoma City AP (S)	35	2	97	4	1280	9	13	100/74	97/74	95/73	23	78	77	76
Ponca City	36	4	97	0	996	5	9	100/74	97/74	94/74	24	77	76	76
Seminole	35	2	96	4	865	11	15	99/74	96/74	94/73	23	77	76	75
Stillwater (S)	36	1	97	1	884	8	13	100/74	96/74	93/74	24	77	76	75
Tulsa AP	36	1	95	5	650	8	13	101/74	98/75	95/75	22	79	78	77
Woodward	36	3	99	3	1900	6	10	100/73	97/73	94/73	26	78	76	75
OREGON														
Albany	44	4	123	1	224	18	22	92/67	89/66	86/65	31	69	67	66
Astoria AP (S)	46	1	123	5	8	25	29	75/65	71/62	68/61	16	65	63	62
Baker AP	44	5	117	5	3368	−1	6	92/63	89/61	86/60	30	65	63	61
Bend	44	0	121	2	3599	−3	4	90/62	87/60	84/59	33	64	62	60
Corvallis (S)	44	3	123	2	221	18	22	92/67	89/66	86/65	31	69	67	66
Eugene AP	44	1	123	1	364	17	22	92/67	89/66	86/65	31	69	67	66
Grants Pass	42	3	123	2	925	20	24	99/69	96/68	93/67	33	71	69	68
Klamath Falls AP	42	1	121	4	4091	4	9	90/61	87/60	84/59	36	63	61	60
Medford AP (S)	42	2	122	5	1298	19	23	98/68	94/67	91/66	35	70	68	67
Pendleton AP	45	4	118	5	1492	−2	5	97/65	93/64	90/62	29	66	65	63
Portland AP	45	4	122	4	21	17	23	89/68	85/67	81/65	23	69	67	66
Portland CO	45	3	122	4	57	18	24	90/68	86/67	82/65	21	69	67	66
Roseburg AP	43	1	123	2	505	18	23	93/67	90/66	87/65	30	69	67	66
Salem AP	45	0	123	0	195	18	23	92/68	88/66	84/65	31	69	68	66
The Dalles	45	4	121	1	102	13	19	93/69	89/68	85/66	28	70	68	67
PENNSYLVANIA														
Allentown AP	40	4	75	3	376	4	9	92/73	88/72	86/72	22	76	75	73
Altoona CO	40	2	78	2	1468	0	5	90/72	87/71	84/70	23	74	73	72
Butler	40	4	80	0	1100	1	6	90/73	87/72	85/71	22	75	74	73
Chambersburg	40	0	77	4	640	4	8	93/75	90/74	87/73	23	77	76	75
Erie AP	42	1	80	1	732	4	9	88/73	85/72	83/71	18	75	74	72

Weather Data and Design Conditions

TABLE 1 CLIMATIC CONDITIONS FOR THE UNITED STATES[a]

Col. 1	Col. 2		Col. 3		Col. 4	Winter,[d] F Col. 5		Summer,[e] F Col. 6			Col. 7	Col. 8		
State and Station	Latitude[b]		Longitude[c]		Elevation[c]	Design Dry-Bulb		Design Dry-Bulb and Mean Coincident Wet-Bulb			Mean Daily	Design Wet-Bulb		
	°	′	°	′	Ft	99%	97.5%	1%	2.5%	5%	Range	1%	2.5%	5%
Harrisburg AP	40	1	76	5	335	7	11	94/75	91/74	88/73	21	77	76	75
Johnstown	40	2	78	5	1214	−3	2	86/70	83/70	80/68	23	72	71	70
Lancaster	40	1	76	2	255	4	8	93/75	90/74	87/73	22	77	76	75
Meadville	41	4	80	1	1065	0	4	88/71	85/70	83/69	21	73	72	71
New Castle	41	0	80	2	825	2	7	91/73	88/72	86/71	23	75	74	73
Philadelphia AP	39	5	75	2	7	10	14	93/75	90/74	87/72	21	77	76	75
Pittsburgh AP	40	3	80	1	1137	1	5	89/72	86/71	84/70	22	74	73	72
Pittsburgh CO	40	3	80	0	749r	3	7	91/72	88/71	86/70	19	74	73	72
Reading CO	40	2	76	0	226	9	13	92/73	89/72	86/72	19	76	75	73
Scranton/ Wilkes-Barre	41	2	75	4	940	1	5	90/72	87/71	84/70	19	74	73	72
State College (S)	40	5	77	5	1175	3	7	90/72	87/71	84/70	23	74	73	72
Sunbury	40	5	76	5	480	2	7	92/73	89/72	86/70	22	75	74	73
Uniontown	39	5	79	4	1040	5	9	91/74	88/73	85/72	22	76	75	74
Warren	41	5	79	1	1280	−2	4	89/71	86/71	83/70	24	74	73	72
West Chester	40	0	75	4	440	9	13	92/75	89/74	86/72	20	77	76	75
Williamsport AP	41	1	77	0	527	2	7	92/73	89/72	86/70	23	75	74	73
York	40	0	76	4	390	8	12	94/75	91/74	88/73	22	77	76	75
RHODE ISLAND														
Newport (S)	41	3	71	2	20	5	9	88/73	85/72	82/70	16	76	75	73
Providence AP	41	4	71	3	55	5	9	89/73	86/72	83/70	19	75	74	73
SOUTH CAROLINA														
Anderson	34	3	82	4	764	19	23	94/74	92/74	90/74	21	77	76	75
Charleston AFB (S)	32	5	80	0	41	24	27	93/78	91/78	89/77	18	81	80	79
Charleston CO	32	5	80	0	9	25	28	94/78	92/78	90/77	13	81	80	79
Columbia AP	34	0	81	1	217	20	24	97/76	95/75	93/75	22	79	78	77
Florence AP	34	1	79	4	146	22	25	94/77	92/77	90/76	21	80	79	78
Georgetown	33	2	79	2	14	23	26	92/79	90/78	88/77	18	81	80	79
Greenville AP	34	5	82	1	957	18	22	93/74	91/74	89/74	21	77	76	75
Greenwood	34	1	82	1	671	18	22	95/75	93/74	91/74	21	78	77	76
Orangeburg	33	3	80	5	244	20	24	97/76	95/75	93/75	20	79	78	77
Rock Hill	35	0	81	0	470	19	23	96/75	94/74	92/74	20	78	77	76
Spartanburg AP	35	0	82	0	816	18	22	93/74	91/74	89/74	20	77	76	75
Sumter-Shaw AFB	34	0	80	3	291	22	25	95/77	92/76	90/75	21	79	78	77
SOUTH DAKOTA														
Aberdeen AP	45	3	98	3	1296	−19	−15	94/73	91/72	88/70	27	77	75	73
Brookings	44	2	96	5	1642	−17	−13	95/73	92/72	89/71	25	77	75	73
Huron AP	44	3	98	1	1282	−18	−14	96/73	93/72	90/71	28	77	75	73
Mitchell	43	5	98	0	1346	−15	−10	96/72	93/71	90/70	28	76	75	73
Pierre AP	44	2	100	2	1718r	−15	−10	99/71	95/71	92/69	29	75	74	72
Rapid City AP (S)	44	0	103	0	3165	−11	−7	95/66	92/65	89/65	28	71	69	67
Sioux Falls AP	43	4	96	4	1420	−15	−11	94/73	91/72	88/71	24	76	75	73
Watertown AP	45	0	97	0	1746	−19	−15	94/73	91/72	88/71	26	76	75	73
Yankton	43	0	97	2	1280	−13	−7	94/73	91/72	88/71	25	77	76	74
TENNESSEE														
Athens	33	3	84	4	940	13	18	95/74	92/73	90/73	22	77	76	75
Bristol- Tri City AP	36	3	82	2	1519	9	14	91/72	89/72	87/71	22	75	75	73
Chattanooga AP	35	0	85	1	670	13	18	96/75	93/74	91/74	22	78	77	76
Clarksville	36	4	87	2	470	6	12	95/76	93/74	90/74	21	78	77	76
Columbia	35	4	87	0	690	10	15	97/75	94/74	91/74	21	78	77	76
Dyersburg	36	0	89	3	334	10	15	96/78	94/77	91/76	21	81	80	78
Greenville	35	5	82	5	1320	11	16	92/73	90/72	88/72	22	76	75	74
Jackson AP	35	4	88	5	413	11	16	98/76	95/75	92/75	21	79	78	77
Knoxville AP	35	5	84	0	980	13	19	94/74	92/73	90/73	21	77	76	75
Memphis AP	35	0	90	0	263	13	18	98/77	95/76	93/76	21	80	79	78
Murfreesboro	35	5	86	2	608	9	14	97/75	94/74	91/74	22	78	77	76
Nashville AP (S)	36	1	86	4	577	9	14	97/75	94/74	91/74	21	78	77	76
Tullahoma	35	2	86	1	1075	8	13	96/74	93/73	91/73	22	77	76	75
TEXAS														
Abilene AP	32	3	99	4	1759	15	20	101/71	99/71	97/71	22	75	74	74
Alice AP	27	4	98	0	180	31	34	100/78	98/77	95/77	20	82	81	79
Amarillo AP	35	1	101	4	3607	6	11	98/67	95/67	93/67	26	71	70	70
Austin AP	30	2	97	4	597	24	28	100/74	98/74	97/74	22	78	77	77
Bay City	29	0	96	0	52	29	33	96/77	94/77	92/77	16	80	79	79
Beaumont	30	0	94	0	18	27	31	95/79	93/78	91/78	19	81	80	80
Beeville	28	2	97	4	225	30	33	99/78	97/77	95/77	18	82	81	79
Big Spring AP (S)	32	2	101	3	2537	16	20	100/69	97/69	95/69	26	74	73	72
Brownsville AP (S)	25	5	97	3	16	35	39	94/77	93/77	92/77	18	80	79	79

TABLE 1 CLIMATIC CONDITIONS FOR THE UNITED STATES[a]

						Winter,[d] F		Summer,[e] F						
Col. 1	Col. 2		Col. 3		Col. 4	Col. 5		Col. 6			Col. 7	Col. 8		
State and Station	Latitude[b]		Longitude[c]		Elevation[c]	Design Dry-Bulb		Design Dry-Bulb and Mean Coincident Wet-Bulb			Mean Daily	Design Wet-Bulb		
	°	'	°	'	Ft	99%	97.5%	1%	2.5%	5%	Range	1%	2.5%	5%
Brownwood	31	5	99	0	1435	18	22	101/73	99/73	96/73	22	77	76	75
Bryan AP	30	4	96	2	275	24	29	98/76	96/76	94/76	20	79	78	78
Corpus Christi AP	27	5	97	3	43	31	35	95/78	94/78	92/78	19	80	80	79
Corsicana	32	0	96	3	425	20	25	100/75	98/75	96/75	21	79	78	77
Dallas AP	32	5	96	5	481	18	22	102/75	100/75	97/75	20	78	78	77
Del Rio, Laughlin AFB	29	2	101	0	1072	26	31	100/73	98/73	97/73	24	79	77	76
Denton	33	1	97	1	655	17	22	101/74	99/74	97/74	22	78	77	76
Eagle Pass	28	5	100	3	743	27	32	101/73	99/73	98/73	24	78	78	77
El Paso AP (S)	31	5	106	2	3918	20	24	100/64	98/64	96/64	27	69	68	68
Fort Worth AP (S)	32	5	97	0	544r	17	22	101/74	99/74	97/74	22	78	77	76
Galveston AP	29	2	94	5	5	31	36	90/79	89/79	88/78	10	81	80	80
Greenville	33	0	96	1	575	17	22	101/74	99/74	97/74	21	78	77	76
Harlingen	26	1	97	4	37	35	39	96/77	94/77	93/77	19	80	79	79
Houston AP	29	4	95	2	50	27	32	96/77	94/77	92/77	18	80	79	79
Houston CO	29	5	95	2	158r	28	33	97/77	95/77	93/77	18	80	79	79
Huntsville	30	4	95	3	494	22	27	100/75	98/75	96/75	20	78	78	77
Killeen-Gray AFB	31	0	97	4	1021	20	25	99/73	97/73	95/73	22	77	76	75
Lamesa	32	5	102	0	2965	13	17	99/69	96/69	94/69	26	73	72	71
Laredo AFB	27	3	99	3	503	32	36	102/73	101/73	99/74	23	78	78	77
Longview	32	2	94	4	345	19	24	99/76	97/76	95/76	20	80	79	78
Lubbock AP	33	4	101	5	3243	10	15	98/69	96/69	94/69	26	73	72	71
Lufkin AP	31	1	94	5	286	25	29	99/76	97/76	94/76	20	80	79	78
Mc Allen	26	1	98	1	122	35	39	97/77	95/77	94/77	21	80	79	79
Midland AP (S)	32	0	102	1	2815r	16	21	100/69	98/69	96/69	26	73	72	71
Mineral Wells AP	32	5	98	0	934	17	22	101/74	99/74	97/74	22	78	77	76
Palestine CO	31	5	95	4	580	23	27	100/76	98/76	96/76	20	79	79	78
Pampa	35	3	101	0	3230	7	12	99/67	96/67	94/67	26	71	70	70
Pecos	31	2	103	3	2580	16	21	100/69	98/69	96/69	27	73	72	71
Plainview	34	1	101	4	3400	8	13	98/68	96/68	94/68	26	72	71	70
Port Arthur AP	30	0	94	0	16	27	31	95/79	93/78	91/78	19	81	80	80
San Angelo, Goodfellow AFB	31	2	100	2	1878	18	22	101/71	99/71	97/70	24	75	74	73
San Antonio AP (S)	29	3	98	3	792	25	30	99/72	97/73	96/73	19	77	76	76
Sherman Perrin AFB	33	4	96	4	763	15	20	100/75	98/75	95/74	22	78	77	76
Snyder	32	4	101	0	2325	13	18	100/70	98/70	96/70	26	74	73	72
Temple	31	1	97	2	675	22	27	100/74	99/74	97/74	22	78	77	77
Tyler AP	32	2	95	2	527	19	24	99/76	97/76	95/76	21	80	79	78
Vernon	34	1	99	2	1225	13	17	102/73	100/73	97/73	24	77	76	75
Victoria AP	28	5	97	0	104	29	32	98/78	96/77	94/77	18	82	81	79
Waco AP	31	4	97	0	500	21	26	101/75	99/75	97/75	22	78	78	77
Wichita Falls AP	34	0	98	3	994	14	18	103/73	101/73	98/73	24	77	76	75
UTAH														
Cedar City AP	37	4	113	1	5613	−2	5	93/60	91/60	89/59	32	65	63	62
Logan	41	4	111	5	4775	−3	2	93/62	91/61	88/60	33	65	64	63
Moab	38	5	109	3	3965	6	11	100/60	98/60	96/60	30	65	64	63
Ogden AP	41	1	112	0	4455	1	5	93/63	91/61	88/61	33	66	65	64
Price	39	4	110	5	5580	−2	5	93/60	91/60	89/59	33	65	63	62
Provo	40	1	111	4	4470	1	6	98/62	96/62	94/61	32	66	65	64
Richfield	38	5	112	0	5300	−2	5	93/60	91/60	89/59	34	65	63	62
St George CO	37	1	113	4	2899	14	21	103/65	101/65	99/64	33	70	68	67
Salt Lake City AP (S)	40	5	112	0	4220	3	8	97/62	95/62	92/61	32	66	65	64
Vernal AP	40	3	109	3	5280	−5	0	91/61	89/60	86/59	32	64	63	62
VERMONT														
Barre	44	1	72	3	1120	−16	−11	84/71	81/69	78/68	23	73	71	70
Burlington AP (S)	44	3	73	1	331	−12	−7	88/72	85/70	82/69	23	74	72	71
Rutland	43	3	73	0	620	−13	−8	87/72	84/70	81/69	23	74	72	71
VIRGINIA														
Charlottsville	38	1	78	3	870	14	18	94/74	91/74	88/73	23	77	76	75
Danville AP	36	3	79	2	590	14	16	94/74	92/73	90/73	21	77	76	75
Fredericksburg	38	2	77	3	50	10	14	96/76	93/75	90/74	21	78	77	76
Harrisonburg	38	3	78	5	1340	12	16	93/72	91/72	88/71	23	75	74	73
Lynchburg AP	37	2	79	1	947	12	16	93/74	90/74	88/73	21	77	76	75
Norfolk AP	36	5	76	1	26	20	22	93/77	91/76	89/76	18	79	78	77
Petersburg	37	1	77	3	194	14	17	95/76	92/76	90/75	20	79	78	77
Richmond AP	37	3	77	2	162	14	17	95/76	92/76	90/75	21	79	78	77

Weather Data and Design Conditions

TABLE 1 CLIMATIC CONDITIONS FOR THE UNITED STATES[a]

						Winter,[d] F		Summer,[e] F						
Col. 1	Col. 2		Col. 3		Col. 4	Col. 5		Col. 6			Col. 7	Col. 8		
State and Station	Latitude[b]		Longitude[c]		Elevation[c]	Design Dry-Bulb		Design Dry-Bulb and Mean Coincident Wet-Bulb			Mean Daily	Design Wet-Bulb		
	°	′	°	′	Ft	99%	97.5%	1%	2.5%	5%	Range	1%	2.5%	5%
Roanoke AP	37	2	80	0	1174r	12	16	93/72	91/72	88/71	23	75	74	73
Staunton	38	2	78	5	1480	12	16	93/72	91/72	88/71	23	75	74	73
Winchester	39	1	78	1	750	6	10	93/75	90/74	88/74	21	77	76	75
WASHINGTON														
Aberdeen	47	0	123	5	12	25	28	80/65	77/62	73/61	16	65	63	62
Bellingham AP	48	5	122	3	150	10	15	81/67	77/65	74/63	19	68	65	63
Bremerton	47	3	122	4	162	21	25	82/65	78/64	75/62	20	66	64	63
Ellensburg AP	47	0	120	3	1729	2	6	94/65	91/64	87/62	34	66	65	63
Everett-Paine AFB	47	5	122	2	598	21	25	80/65	76/64	73/62	20	67	64	63
Kennewick	46	0	119	1	392	5	11	99/68	96/67	92/66	30	70	68	67
Longview	46	1	123	0	12	19	24	88/68	85/67	81/65	30	69	67	66
Moses Lake, Larson AFB	47	1	119	2	1183	1	7	97/66	94/65	90/63	32	67	66	64
Olympia AP	47	0	122	5	190	16	22	87/66	83/65	79/64	32	67	66	64
Port Angeles	48	1	123	3	99	24	27	72/62	69/61	67/60	18	64	62	61
Seattle-Boeing Field	47	3	122	2	14	21	26	84/68	81/66	77/65	24	69	67	65
Seattle CO (S)	47	4	122	2	14	22	27	85/68	82/66	78/65	19	69	67	65
Seattle-Tacoma AP (S)	47	3	122	2	386	21	26	84/65	80/64	76/62	22	66	64	63
Spokane AP (S)	47	4	117	3	2357	−6	2	93/64	90/63	87/62	28	65	64	62
Tacoma-Mc Chord AFB	47	1	122	3	350	19	24	86/66	82/65	79/63	22	68	66	64
Walla Walla AP	46	1	118	2	1185	0	7	97/67	94/66	90/65	27	69	67	66
Wenatchee	47	2	120	2	634	7	11	99/67	96/66	92/64	32	68	67	65
Yakima AP	46	3	120	3	1061	−2	5	96/65	93/65	89/63	36	68	66	65
WEST VIRGINIA														
Beckley	37	5	81	1	2330	−2	4	83/71	81/69	79/69	22	73	71	70
Bluefield AP	37	2	81	2	2850	−2	4	83/71	81/69	79/69	22	73	71	70
Charleston AP	38	2	81	4	939	7	11	92/74	90/73	87/72	20	76	75	74
Clarksburg	39	2	80	2	977	6	10	92/74	90/73	87/72	21	76	75	74
Elkins AP	38	5	79	5	1970	1	6	86/72	84/70	82/70	22	74	72	71
Huntington CO	38	2	82	3	565r	5	10	94/76	91/74	89/73	22	78	77	75
Martinsburg AP	39	2	78	0	537	6	10	93/75	90/74	88/74	21	77	76	75
Morgantown AP	39	4	80	0	1245	4	8	90/74	87/73	85/73	22	76	75	74
Parkersburg CO	39	2	81	3	615r	7	11	93/75	90/74	88/73	21	77	76	75
Wheeling	40	1	80	4	659	1	5	89/72	86/71	84/70	21	74	73	72
WISCONSIN														
Appleton	44	2	88	2	742	−14	−9	89/74	86/72	83/71	23	76	74	72
Ashland	46	3	90	5	650	−21	−16	85/70	82/68	79/66	23	72	70	68
Beloit	42	3	89	0	780	−7	−3	92/75	90/75	88/74	24	78	77	75
Eau Claire AP	44	5	91	3	888	−15	−11	92/75	89/73	86/71	23	77	75	73
Fond du Lac	43	5	88	3	760	−12	−8	89/74	86/72	84/71	23	76	74	72
Green Bay AP	44	3	88	1	683	−13	−9	88/74	85/72	83/71	23	76	74	72
La Crosse AP	43	5	91	2	652	−13	−9	91/75	88/73	85/72	22	77	75	74
Madison AP (S)	43	1	89	2	858	−11	−7	91/74	88/73	85/71	22	77	75	73
Manitowoc	44	1	87	4	660	−11	−7	89/74	86/72	83/71	21	76	74	72
Marinette	45	0			605	−15	−11	87/73	84/71	82/70	20	75	73	71
Milwaukee AP	43	0	87	5	672	−8	−4	90/74	87/73	84/71	21	76	74	73
Racine	42	4	87	4	640	−6	−2	91/75	88/73	85/72	21	77	75	74
Sheboygan	43	4	87	4	648	−10	−6	89/75	86/73	83/72	20	77	75	74
Stevens Point	43	0	89	3	1079	−15	−11	92/75	89/73	86/71	23	77	75	73
Waukesha	43	0	88	1	860	−9	−5	90/74	87/73	84/71	22	76	74	73
Wausau AP	44	6	89	4	1196	−16	−12	91/74	88/72	85/70	23	76	74	72
WYOMING														
Casper AP	42	5	106	3	5319	−11	−5	92/58	90/57	87/57	31	63	61	60
Cheyene AP	41	1	104	5	6126	−9	−1	89/58	86/58	84/57	30	63	62	60
Cody AP	44	3	109	0	5090	−19	−13	89/60	86/60	83/59	32	64	63	61
Evanston	41	2	111	0	6860	−9	−3	86/55	84/55	82/54	32	59	58	57
Lander AP (S)	42	5	108	4	5563	−16	−11	91/61	88/61	85/60	32	64	63	61
Laramie AP (S)	41	2	105	3	7266	−14	−6	84/56	81/56	79/55	28	61	60	59
Newcastle	43	5	104	1	4480	−17	−12	91/64	87/63	84/63	30	69	68	66
Rawlins	41	5	107	1	6736	−12	−4	86/57	83/57	81/56	40	62	61	60
Rock Springs AP	41	4	109	0	6741	−9	−3	86/55	84/55	82/54	32	59	58	57
Sheridan AP	44	5	107	0	3942	−14	−8	94/62	91/62	88/61	32	66	65	63
Torrington	4	0	104	1	4098	−14	−8	94/62	91/62	88/61	30	66	65	63

CHAPTER 24 **1985 Fundamentals Handbook**

TABLE 2 CLIMATIC CONDITIONS FOR CANADA[a]

Col. 1	Col. 2		Col. 3.		Col. 4.	WINTER[e], F Col. 5.		SUMMER[f], F Col. 6			Col. 7	Col. 8		
Province and Station[b]	Latitude[c]		Longitude[c]		Elevation[d]	Design Dry-Bulb		Design Dry Bulb and Mean Coincident Wet-Bulb			Mean Daily	Design Wet Bulb		
	°	′	°	′	Ft	99%	97.5%	1%	2.5%	5%	Range	1%	2.5%	5%
ALBERTA														
Calgary AP	51	6	114	1	3540	−27	−23	84/63	81/61	79/60	25	65	63	62
Edmonton AP	53	34	113	31	2219	−29	−25	85/66	82/65	79/63	23	68	66	65
Grande Prairie AP	55	11	118	53	2190	−39	−33	83/64	80/63	78/61	23	66	64	62
Jasper	52	53	118	4	3480	−31	−26	83/64	80/62	77/61	28	66	64	63
Lethbridge AP (S)	49	38	112	48	3018	−27	−22	90/65	87/64	84/63	28	68	66	65
McMurray AP	56	39	111	13	1216	−41	−38	86/67	82/65	79/64	26	69	67	65
Medicine Hat AP	50	1	110	43	2365	−29	−24	93/66	90/65	87/64	28	70	68	66
Red Deer AP	52	11	113	54	2965	−31	−26	84/65	81/64	78/62	25	67	66	64
BRITISH COLUMBIA														
Dawson Creek	55	44	120	11	2164	−37	−33	82/64	79/63	76/61	26	66	64	62
Fort Nelson AP (S)	58	50	122	35	1230	−43	−40	84/64	81/63	78/62	23	67	65	64
Kamloops CO	50	43	120	25	1133	−21	−15	94/66	91/65	88/64	29	68	66	65
Nanaimo (S)	49	11	123	58	230	16	20	83/67	80/65	77/64	21	68	66	65
New Westminster	49	13	122	54	50	14	18	84/68	81/67	78/66	19	69	68	66
Penticton AP	49	28	119	36	1121	0	4	92/68	89/67	87/66	31	70	68	67
Prince George AP(S)	53	53	122	41	2218	−33	−28	84/64	80/62	77/61	26	66	64	62
Prince Rupert CO	54	17	130	23	170	− 2	2	64/59	63/57	61/56	12	60	58	57
Trail	49	8	117	44	1400	− 5	0	92/66	89/65	86/64	33	68	67	65
Vancouver AP (S)	49	11	123	10	16	15	19	79/67	77/66	74/65	17	68	67	66
Victoria CO	48	25	123	19	228	20	23	77/64	73/62	70/60	16	64	62	60
MANITOBA														
Brandon	49	52	99	59	1200	−30	−27	89/72	86/70	83/68	25	74	72	70
Churchill AP (S)	58	45	94	4	155	−41	−39	81/66	77/64	74/62	18	67	65	63
Dauphin AP	51	6	100	3	999	−31	−28	87/71	84/70	81/68	23	74	72	70
Flin Flon	54	46	101	51	1098	−41	−37	84/68	81/66	79/65	19	70	68	67
Portage la Prairie AP	49	54	98	16	867	−28	−24	88/73	86/72	83/70	22	76	74	71
The Pas AP (S)	53	58	101	6	894	−37	−33	85/68	82/67	79/66	20	71	69	68
Winnipeg AP (S)	49	54	97	14	786	−30	−27	89/73	86/71	84/70	22	75	73	71
NEW BRUNSWICK														
Campbellton CO	48	0	66	40	25	−18	−14	85/68	82/67	79/66	21	72	70	68
Chatham AP	47	1	65	27	112	−15	−10	89/69	85/68	82/67	22	72	71	69
Edmundston CO	47	22	68	20	500	−21	−16	87/70	83/68	80/67	21	73	71	69
Fredericton AP (S)	45	52	66	32	74	−16	−11	89/71	85/69	82/68	23	73	71	70
Moncton AP (S)	46	7	64	41	248	−12	− 8	85/70	82/69	79/67	23	72	71	69
Saint John AP	45	19	65	53	352	−12	− 8	80/67	77/65	75/64	19	70	68	66
NEWFOUNDLAND														
Corner Brook	48	58	57	57	15	− 5	0	76/64	73/63	71/62	17	67	66	65
Gander AP	48	57	54	34	482	− 5	− 1	82/66	79/65	77/64	19	69	67	66
Goose Bay AP (S)	53	19	60	25	144	−27	−24	85/66	81/64	77/63	19	68	66	64
St. John's AP (S)	47	37	52	45	463	3	7	77/66	75/65	73/64	18	69	67	66
Stephenville AP	48	32	58	33	44	− 3	4	76/65	74/64	71/63	14	67	66	65
NORTHWEST TERR.														
Fort Smith AP (S)	60	1	111	58	665	−49	−45	85/66	81/64	78/63	24	68	66	65
Frobisher AP (S)	63	45	68	33	68	−43	−41	66/53	63/51	59/50	14	54	52	51
Inuvik (S)	68	18	133	29	200	−56	−53	79/62	77/60	75/59	21	64	62	61
Resolute AP (S)	74	43	94	59	209	−50	−47	57/48	54/46	51/45	10	50	48	46
Yellowknife AP	62	28	114	27	682	−49	−46	79/62	77/61	74/60	16	64	63	62
NOVA SCOTIA														
Amherst	45	49	64	13	65	−11	− 6	84/69	81/68	79/67	21	72	70	68
Halifax AP (S)	44	39	63	34	83	1	5	79/66	76/65	74/64	16	69	67	66
Kentville (S)	45	3	64	36	40	− 3	1	85/69	83/68	80/67	22	72	71	69
New Glasgow	45	37	62	37	317	− 9	− 5	81/69	79/68	77/67	20	72	70	69
Sydney AP	46	10	60	3	197	− 1	3	82/69	80/68	77/66	19	71	70	68
Truro CO	45	22	63	16	131	− 8	− 5	82/70	80/69	78/68	22	73	71	70
Yarmouth AP	43	50	66	5	136	5	9	74/65	72/64	70/63	15	68	66	65

[a] Table 2 was prepared by ASHRAE Technical Committee 4.2 using data compiled from official weather stations where hourly weather observations are made by trained observers.
[b] When airport temperature observations were used to develop design data, "AP" follows the station name. Data for stations followed by "CO" come from office locations within an urban area and generally reflect an influence of the surroundings area. Stations without designation can be considered semirural and can be directly compared to most airport data.
[c] Latitude, for use in calculating solar loads, and longitude are given to the nearest minute.
[d] Elevations are ground elevations for each station as of 1964. Temperature readings are generally made at an elevation of 5 ft above ground.
[e] The winter design data are based on the month of January only.
[f] The summer design data are based on the month of July only. See also Ref 18.

Weather Data and Design Conditions

TABLE 2 CLIMATIC CONDITIONS FOR CANADA[a]

Col. 1	Col. 2		Col. 3.		Col. 4.	WINTER[e], F Col. 5.		SUMMER[f], F Col. 6			Col. 7	Col. 8		
Province and Station[b]	Latitude[c]		Longitude[c]		Elevation[d]	Design Dry-Bulb		Design Dry Bulb and Mean Coincident Wet-Bulb			Mean Daily Range	Design Wet Bulb		
	°	′	°	′	Ft	99%	97.5%	1%	2.5%	5%		1%	2.5%	5%
ONTARIO														
Belleville	44	9	77	24	250	−11	− 7	86/73	84/72	82/71	20	75	74	73
Chatham	42	24	82	12	600	0	3	89/74	87/73	85/72	19	76	75	74
Cornwall	45	1	74	45	210	−13	− 9	89/73	87/72	84/71	21	75	74	72
Hamilton	43	16	79	54	303	− 3	1	88/73	86/72	83/71	21	76	74	73
Kapuskasing AP (S)	49	25	82	28	752	−31	−28	86/70	83/69	80/67	23	72	70	69
Kenora AP	49	48	94	22	1345	−32	−28	84/70	82/69	80/68	19	73	71	70
Kingston	44	16	76	30	300	−11	− 7	87/73	84/72	82/71	20	75	74	73
Kitchener	43	26	80	30	1125	− 6	− 2	88/73	85/72	83/71	23	75	74	72
London AP	43	2	81	9	912	− 4	0	87/74	85/73	83/72	21	76	74	73
North Bay AP	46	22	79	25	1210	−22	−18	84/68	81/67	79/66	20	71	70	68
Oshawa	43	54	78	52	370	− 6	− 3	88/73	86/72	84/71	20	75	74	73
Ottawa AP (S)	45	19	75	40	413	−17	−13	90/72	87/71	84/70	21	75	73	72
Owen Sound	44	34	80	55	597	− 6	− 2	84/71	82/70	80/69	21	73	72	70
Peterborough	44	17	78	19	635	−13	− 9	87/72	85/71	83/70	21	75	73	72
St. Catharines	43	11	79	14	325	− 1	3	87/73	85/72	83/71	20	76	74	73
Sarnia	42	58	82	22	625	0	3	88/73	86/72	84/71	19	76	74	73
Sault Ste. Marie AP	46	32	84	30	675	−17	−13	85/71	82/69	79/68	22	73	71	70
Sudbury AP	46	37	80	48	1121	−22	−19	86/69	83/67	81/66	22	72	70	68
Thunder Bay AP	48	22	89	19	644	−27	−24	85/70	83/68	80/67	24	72	70	68
Timmins AP	48	34	81	22	965	−33	−29	87/69	84/68	81/66	25	72	70	68
Toronto AP (S)	43	41	79	38	578	− 5	− 1	90/73	87/72	85/71	20	75	74	73
Windsor AP	42	16	82	58	637	0	4	90/74	88/73	86/72	20	77	75	74
PRINCE EDWARD ISLAND														
Charlottetown AP (S)	46	17	63	8	186	− 7	− 4	80/69	78/68	76/67	16	71	70	68
Summerside AP	46	26	63	50	78	− 8	− 4	81/69	79/68	77/67	16	72	70	68
QUEBEC														
BagotvilleAP	48	20	71	0	536	−28	−23	87/70	83/68	80/67	21	72	70	68
Chicoutimi	48	25	71	5	150	−26	−22	86/70	83/68	80/67	20	72	70	68
Drummondville	45	53	72	29	270	−18	−14	88/72	85/71	82/69	21	75	73	71
Granby	45	23	72	42	550	−19	−14	88/72	85/71	83/70	21	75	73	72
Hull	45	26	75	44	200	−18	−14	90/72	87/71	84/70	21	75	73	72
Megantic AP	45	35	70	52	1362	−20	−16	86/71	83/70	81/69	20	74	72	71
Montreal AP (S)	45	28	73	45	98	−16	−10	88/73	85/72	83/71	17	75	74	72
Quebec AP	46	48	71	23	245	−19	−14	87/72	84/70	81/68	20	74	72	70
Rimouski	48	27	68	32	117	−16	−12	83/68	79/66	76/65	18	71	69	67
St. Jean	45	18	73	16	129	−15	−11	88/73	86/72	84/71	20	75	74	72
St. Jerome	45	48	74	1	556	−17	−13	88/72	86/71	83/70	23	75	73	72
Sept. Iles AP (S)	50	13	66	16	190	−26	−21	76/63	73/61	70/60	17	67	65	63
Shawinigan	46	34	72	43	306	−18	−14	86/72	84/70	82/69	21	74	72	71
Sherbrooke CO	45	24	71	54	595	−25	−21	86/72	84/71	81/69	20	74	73	71
Thetford Mines	46	4	71	19	1020	−19	−14	87/71	84/70	81/69	21	74	72	71
Trois Rivières	46	21	72	35	50	−17	−13	88/72	85/70	82/69	23	74	72	71
Val d'Or AP	48	3	77	47	1108	−32	−27	85/70	83/68	80/67	22	72	70	68
Valleyfield	45	16	74	6	150	−14	−10	89/73	86/72	84/71	20	75	74	72
SASKATCHEWAN														
Estevan AP	49	4	103	0	1884	−30	−25	92/70	89/68	86/67	26	72	70	69
Moose Jaw AP	50	20	105	33	1857	−29	−25	93/69	89/67	86/66	27	71	69	68
North Battleford AP	52	46	108	15	1796	−33	−30	88/67	85/66	82/65	23	69	68	66
Prince Albert AP	53	13	105	41	1414	−42	−35	87/67	84/66	81/65	25	70	68	67
Regina AP	50	26	104	40	1884	−33	−29	91/69	88/68	84/67	26	72	70	68
Saskatoon AP (S)	52	10	106	41	1645	−35	−31	89/68	86/66	83/65	26	70	68	67
Swift Current AP (S)	50	17	107	41	2677	−28	−25	93/68	90/66	87/65	25	70	69	67
Yorkton AP	51	16	102	28	1653	−35	−30	87/69	84/68	80/66	23	72	70	68
YUKON TERRITORY														
Whitehorse AP (S)	60	43	135	4	2289	−46	−43	80/59	77/58	74/56	22	61	59	58

Appendix: ASHRAE Winter Design Temperatures

Table 1 .... Winter Climatic Conditions[a]

Col. 1 State	Col. 2 Station	Col. 3[b]	Col. 4[c] Elev, Ft.	Col. 5 Design Winter Temperatures			Col. 6[d] Coincident Wind Velocity
				Median of Extremes	99%	97½%	
Ala.	Anniston	AP	599	12	17	19	L
	Birmingham	AP	610	14	19	22	L
	Dothan	AP	321	19	23	27	L
	Florence	AP	578	8	13	17	L
	Huntsville	AP	619	8	13	17	L
	Mobile	AP	211	21	26	29	M
	Mobile	CO	119	24	28	32	M
	Montgomery	AP	195	18	22	26	L
	Tuscaloosa	AP	170r	14	19	23	L
Alaska	Anchorage	AP	90	−29	−25	−20	VL
	Fairbanks	AP	436	−59	−53	−50	VL
	Juneau	AP	17	−11	−7	−4	L
Ariz.†	Flagstaff	AP	6993	−10	0	5	VL
	Fort Huachuca	AP	4664	18	25	28	VL
	Gila Bend	AP	853	24	30	34	VL
	Kingman	AP	3446	18	25	29	VL
	Phoenix	AP	1117	25	31	34	VL
	Prescott	AP	5014	7	15	19	VL
	Tucson	AP	2584	23	29	32	VL
	Winslow	AP	4880	2	9	13	VL
	Yuma	AP	199	32	37	40	VL
Ark.	Blytheville	AFB	264	6	12	17	L
	El Dorado	AP	252	13	19	23	L
	Fayetteville	AP	1253	3	9	13	M
	Fort Smith	AP	449	9	15	19	M
	Little Rock	AP	257	13	19	23	M
	Pine Bluff	AP	204	14	20	24	L
	Texarkana	AP	361	16	22	26	M
Calif.†	Bakersfield	AP	495	26	31	33	VL
	Barstow	AP	2142	18	24	28	VL
	Blythe	AP	390	26	31	35	VL
	Burbank	AP	699	30	36	38	VL
	Crescent City	AP	50	28	33	36	VL
	El Centro	AP	−30	26	31	35	VL
	Eureka/Arcata	AP	203	30	35	38	VL
	Fresno	AP	326	25	28	31	VL
	Hamilton	AFB	3	28	33	35	VL
	Laguna Beach	SDP	35r	32	37	39	VL
	Long Beach	AP	34	31	36	38	VL
	Los Angeles	AP	99	36	41	43	VL
	Los Angeles	CO	312	38	42	44	VL
	Merced, Castle	AFB	178	24	30	32	VL
	Needles	AP	913	27	33	37	VL
	Oakland	AP	3	30	35	37	VL
	Oxnard	AFB	43	32	35	37	VL
	Palmdale	AP	2517	18	24	27	VL
	Palm Springs	FD	411	27	32	36	VL
	Pomona	CO	871	26	31	34	VL
	Redding	AP	495	25	31	35	VL
	Riverside, March	AFB	1511	26	32	34	VL
	Sacramento	AP	17	24	30	32	VL
	Salinas	AP	74	27	32	35	VL
	San Bernardino, Norton	AFB	1125	26	31	33	VL
	San Diego	AP	19r	38	42	44	VL
	San Francisco	AP	8	32	35	37	VL
	San Francisco	CO	52	38	42	44	VL
	San Jose	AP	70r	30	34	36	VL
	Santa Ana	FD	115r	28	33	36	VL
	Santa Barbara	FD	100	30	34	36	VL
	Santa Maria	AP	238	28	32	34	VL
	Santa Monica	CO	57	38	43	45	VL
	Stockton	AP	28	25	30	34	VL
	Ukiah	AP	620	22	27	30	VL
	Yreka/Montague	FS	2625	7	13	17	VL

DP 716

Table 1 . . . . Winter Climatic Conditions (*Continued*)[a]

Col. 1 State	*Col. 2* Station	*Col. 3*[b]	*Col. 4*[c] Elev, Ft.	*Col. 5* Design Winter Temperatures			*Col. 6*[d] Coincident Wind Velocity
				Median of Extremes	99%	97½%	
Colo.	Colorado Springs	AP	6173	−9	−1	4	L
	Denver	AP	5283	−9	−2	3	L
	Grand Junction	AP	4849	−2	8	11	VL
	La Junta	AP	4188	−14	−6	−2	M
	Pueblo	AP	4639	−14	−5	−1	L
	Trinidad	AP	5746	−9	1	5	L
Conn.	Hartford, Brainerd Fld.	AP	15	−4	1	5	M
	New Haven	AP	6	0	5	9	H
Del.	Dover	AFB	38	8	13	15	M
	Wilmington	AP	78	6	12	15	M
D.C.	Washington Natl.	AP	14	12	16	19	M
	Andrews	AFB	279	9	13	16	M
Fla.	Cape Kennedy	AP	16	33	37	40	L
	Cross City	AP	46	20	25	29	L
	Daytona Beach	AP	31	28	32	36	L
	Fort Meyers	AP	13	34	38	42	M
	Jacksonville	AP	24	26	29	32	L
	Miami	AP	7	39	44	47	M
	Orlando	AP	106r	29	33	37	L
	Panama City, Tyndall	AFB	22	28	32	35	M
	Pensacola	CO	13r	25	29	32	M
	Tallahassee	AP	58	21	25	29	L
	Tampa	AP	19	32	36	39	M
	West Palm Beach	AP	15r	36	40	44	M
Ga.	Albany	AP	204	21	26	30	L
	Atlanta	AP	1005	14	18	23	H
	Augusta	AP	143	17	20	23	L
	Columbus, Lawson	AFB	242	19	23	26	L
	Macon	AP	356	18	23	27	L
	Rome	AP	637	11	16	20	L
	Savannah, Travis Fld.	AP	52	21	24	27	L
	Valdosta	AP	200	24	28	31	L
Hawaii	Hilo	AP	31	56	59	61	L
	Honolulu	AP	7r	58	60	62	L
Idaho	Boise	AP	2842	0	4	10	L
	Coeur D'Alene	AP	2973	−4	2	7	VL
	Idaho Falls	AP	4730r	−17	−12	−6	VL
	Lewiston	AP	1413	1	6	12	VL
	Mountain Home	AFB	2992	−3	2	9	L
	Pocatello	AP	4444	−12	−8	−2	VL
	Twin Falls	AP	4148	−5	4	8	L
Ill.	Belleville, Scott	AFB	447	0	6	10	M
	Chicago, Midway	AP	610	−7	−4	1	M
	Chicago, O'Hare	AP	658	−9	−4	0	M
	Chicago	CO	594	−5	−3	1	M
	Joliet	AP	588	−11	−5	−1	M
	Moline	AP	582	−12	−7	−3	M
	Peoria	AP	652	−8	−2	2	M
	Quincy	AP	762	−8	−2	2	M
	Rantoul, Chanute	AFB	740	−7	−1	3	M
	Springfield	AP	587	−7	−1	4	M
Ind.	Columbus, Bakalar	AFB	661	−3	3	7	M
	Evansville	AP	381	1	6	10	M
	Fort Wayne	AP	791	−5	0	5	M
	Indianapolis	AP	793	−5	0	4	M
	South Bend	AP	773	−6	−2	3	M
	Terre Haute	AP	601	−3	3	7	M
Iowa	Burlington	AP	694	−10	−4	0	M
	Cedar Rapids	AP	863	−14	−8	−4	M
	Des Moines	AP	948r	−13	−7	−3	M
	Mason City	AP	1194	−20	−13	−9	M
	Ottumwa	AP	842	−12	−6	−2	M
	Sioux City	AP	1095	−17	−10	−6	M

Table 1 .... Winter Climatic Conditions (*Continued*)[a]

Col. 1 *State*	*Col. 2* *Station*	*Col. 3*[b]	*Col. 4*[c] *Elev, Ft.*	*Col. 5* *Design Winter Temperatures*			*Col. 6*[d] *Coincident Wind Velocity*
				Median of Extremes	99%	97½%	
Kans.	Chanute	AP	977	−3	3	7	M
	Dodge City	AP	2594	−5	3	7	M
	Garden City	AP	2882	−10	−1	3	M
	Goodland	AP	3645	−10	−2	4	M
	Hutchinson	AP	1524	−5	2	6	H
	Russell	AP	1864	−7	0	4	M
	Topeka	AP	877	−4	3	6	M
	Wichita	AP	1321	−1	5	9	H
Ky.	Bowling Green	AP	535	1	7	11	L
	Corbin	AP	1175	0	5	9	L
	Lexington	AP	979	0	6	10	M
	Louisville	AP	474	1	8	12	L
	Paducah	AP	398	4	10	14	L
La.	Alexandria	AP	92	20	25	29	L
	Baton Rouge	AP	64	22	25	30	L
	Lafayette	AP	38	23	28	32	L
	Lake Charles	AP	14	25	29	33	M
	Monroe	AP	78	18	23	27	L
	New Orleans	AP	3	29	32	35	M
	Shreveport	AP	252	18	22	26	M
Me.	Augusta	AP	350	−13	−7	−3	M
	Bangor, Dow	AFB	202	−14	−8	−4	M
	Caribou	AP	624	−24	−18	−14	L
	Portland	AP	61	−14	−5	0	L
Md.	Baltimore	AP	146	8	12	15	M
	Baltimore	CO	14r	12	16	20	M
Mass.	Boston	AP	15	−1	6	10	H
	Springfield, Westover	AFB	247	−8	−3	4	M
	Worcester	AP	986	−8	−3	1	M
Mich.	Battle Creek	AP	939	−6	1	5	M
	Detroit, Met.	CAP	633	0	4	8	M
	Flint	AP	766	−7	−1	3	M
	Grand Rapids	AP	681	−3	2	6	M
	Lansing	AP	852	−4	2	6	M
	Muskegon	AP	627	−2	4	8	M
	Saginaw	AP	662	−7	−1	3	M
	Sault Ste. Marie	AP	721	−18	−12	−8	L
	Traverse City	AP	618	−6	0	4	M
	Ypsilanti	AP	777	−3	−1	5	M
Minn.	Alexandria	AP	1421	−26	−19	−15	L
	Duluth	AP	1426	−25	−19	−15	M
	Int. Falls	AP	1179	−35	−29	−24	L
	Minneapolis-St. Paul	AP	822	−19	−14	−10	L
	Rochester	AP	1297	−23	−17	−13	M
	St. Cloud	AP	1034	−26	−20	−16	L
Miss.	Biloxi	AP	25	26	30	32	M
	Columbus	AFB	224	13	18	22	L
	Greenville	AFB	139	16	21	24	L
	Jackson	AP	330	17	21	24	L
	Meridian	AP	294	15	20	24	L
Mo.	Columbia	AP	778	−4	2	6	M
	Farmington	AP	928	−2	4	8	M
	Joplin	AP	982	3	5	7	M
	Kansas City	AP	742	−2	4	8	M
	Kirksville	AP	966	−13	−7	−3	M
	St. Joseph	AP	809	−8	−1	3	M
	St. Louis	AP	535	−2	4	8	M
	St. Louis	CO	465	1	7	11	M
	Springfield	AP	1265	0	5	10	M
Mont.	Billings	AP	3567	−19	−10	−6	L
	Butte	AP	5526r	−34	−24	−16	VL
	Cut Bank	AP	3838r	−32	−23	−17	L
	Glasgow	AP	2277	−33	−25	−20	L

Table 1 . . . . Winter Climatic Conditions (*Continued*)[a]

Col. 1 State	Col. 2 Station	Col. 3[b]	Col. 4[c] Elev, Ft.	Col. 5 Design Winter Temperatures: Median of Extremes	99%	97½%	Col. 6[d] Coincident Wind Velocity
Mont. (Cont'd)	Great Falls	AP	3664r	−29	−20	−16	L
	Helena	AP	3893	−27	−17	−13	L
	Kalispell	AP	2965	−17	−7	−3	VL
	Lewiston	AP	4132	−27	−18	−14	L
	Livingston	AP	4653	−26	−17	−13	L
	Miles City	AP	2629	−27	−19	−15	L
	Missoula	AP	3200	−16	−7	−3	VL
Neb.	Chadron	AP	3300	−21	−13	−9	M
	Grand Island	AP	1841	−14	−6	−2	M
	Lincoln	CO	1150	−10	−4	0	M
	North Platte	AP	2779	−13	−6	−2	M
	Omaha	AP	978	−12	−5	−1	M
	Scottsbluff	AP	3950	−16	−8	−4	M
	Sidney	AP	4292	−15	−7	−2	M
Nev.†	Elko	AP	5075	−21	−13	−7	VL
	Ely	AP	6257	−15	−6	−2	VL
	Fallon	NAS	3934	−2	5	9	VL
	Las Vegas	AP	2162	18	23	26	VL
	Lovelock	AP	3900	0	7	11	VL
	Reno	AP	4404	−2	2	7	VL
	Tonopah	AP	5426	2	9	13	VL
	Winnemucca	AP	4299	−8	0	4	VL
N.H.	Concord	AP	339	−17	−11	−7	M
	Manchester	AP	253	−11	−5	1	M
N.J.	Atlantic City	CO	45	10	14	18	H
	Newark	AP	11	6	11	15	M
	Trenton	CO	144	7	12	16	M
N.M.	Alamagordo, Holloman	AFB	4070	12	18	22	L
	Albuquerque	AP	5310	6	14	17	L
	Carlsbad	AP	3234	11	17	21	L
	Clovis	AP	4279	2	14	17	L
	Farmington	AP	5495	−3	6	9	VL
	Hobbs	AP	3664	9	15	19	L
	Roswell	AP	3612r	5	16	19	L
	Santa Fe	CO	7045	−2	7	11	L
	Tucumcari	AP	4053	1	9	13	L
N.Y.	Albany	AP	277	−14	−5	0	L
	Albany	CO	19r	−5	1	5	L
	Binghamton	CO	858	−8	−2	2	L
	Buffalo	AP	705r	−3	3	6	M
	Elmira	AP	860	−5	1	5	L
	Massena	AP	202r	−22	−16	−12	M
	Newburgh, Stewart	AFB	460	−4	2	6	M
	New York, Kennedy	AP	16	12	17	21	H
	New York, LaGuardia	AP	19r	7	12	16	H
	Rochester	AP	543	−5	2	5	M
	Suffolk County	AFB	57	4	9	13	H
	Syracuse	AP	424	−10	−2	2	M
N.C.	Asheville	AP	2170r	8	13	17	L
	Charlotte	AP	735	13	18	22	L
	Elizabeth City	AP	10	14	18	22	M
	Fayetteville, Pope	AFB	95	13	17	20	L
	Greensboro	AP	897	9	14	17	L
	New Bern	AP	17	14	18	22	L
	Raleigh/Durham	AP	433	13	16	20	L
	Wilmington	AP	30	19	23	27	L
	Winston-Salem	AP	967	9	14	17	L
N.D.	Bismarck	AP	1647	−31	−24	−19	VL
	Dickenson	AP	2595	−31	−23	−19	L
	Fargo	AP	900	−28	−22	−17	L
	Grand Forks	AP	832	−30	−26	−23	L

Table 1 .... Winter Climatic Conditions (*Continued*)[a]

Col. 1 State	Col. 2 Station	Col. 3[b]	Col. 4[c] Elev, Ft.	Col. 5 Design Winter Temperatures			Col. 6[d] Coincident Wind Velocity
				Median of Extremes	99%	97½	
N.D. (Cont'd)	Jamestown	AP	1492	−29	−22	−18	L
	Minot	AP	1713	−31	−24	−20	M
Ohio	Akron/Canton	AP	1210	−5	1	6	M
	Cincinnati	AP	869	−3	3	8	M
	Cincinnati	CO	761	2	8	12	L
	Cleveland	AP	777r	−2	2	7	M
	Columbus	AP	812	−1	2	7	M
	Dayton	AP	997	−2	0	6	M
	Findlay	AP	797	−6	0	4	M
	Mansfield	AP	1297	−11	−5	−1	M
	Toledo	AP	676r	−5	1	5	M
	Youngstown	AP	1178	−5	1	6	M
	Zanesville	AP	881	−7	−1	3	M
Okla.	Ardmore	ED	880	9	15	19	H
	Enid, Vance	AFB	1287	3	10	14	H
	Lawton	AP	1108	6	13	16	H
	Muskogee	AP	610	6	12	16	M
	Oklahoma City	AP	1280	4	11	15	H
	Tulsa	AP	650	4	12	16	H
Ore.	Baker	AP	3368	−10	−3	1	VL
	Eugene	AP	364	16	22	26	VL
	Klamath Falls	AP	4091	−5	1	5	VL
	Medford	AP	1298	15	21	23	VL
	Pendleton	AP	1492	−2	3	10	VL
	Portland	AP	21	17	21	24	L
	Portland	CO	57	21	26	29	L
	Salem	AP	195	15	21	25	VL
Pa.	Allentown	AP	376	−2	3	5	M
	Altoona	CO	1468	−4	1	5	L
	Erie	AP	732	1	7	11	M
	Harrisburg	AP	335	4	9	13	L
	Philadelphia	AP	7	7	11	15	M
	Pittsburgh	AP	1137	−1	5	9	M
	Pittsburgh	CO	749r	1	7	11	M
	Wilkes Barre/Scranton	AP	940	−3	2	6	L
	Williamsport	AP	527	−5	1	5	L
R.I.	Providence	AP	55	0	6	10	M
S.C.	Charleston	AFB	41	19	23	27	L
	Columbia	AP	217	16	20	23	L
	Florence	AP	146	16	21	25	L
	Greenville	AP	957	14	19	23	L
	Spartanburg	AP	816	13	18	22	L
	Sumter, Shaw	AFB	291	18	23	26	L
S.D.	Aberdeen	AP	1296	−29	−22	−18	L
	Huron	AP	1282	−24	−16	−12	L
	Pierre	AP	1718r	−21	−13	−9	M
	Rapid City	AP	3165	−17	−9	−6	M
	Sioux Falls	AP	1420	−21	−14	−10	M
	Watertown	AP	1746	−27	−20	−16	L
Tenn.	Bristol	AP	1519	6	11	16	L
	Chattanooga	AP	670	11	15	19	L
	Jackson	AP	413	8	14	17	L
	Knoxville	AP	980	9	13	17	L
	Memphis	AP	263	11	17	21	L
	Nashville	AP	577	6	12	16	L
Texas	Abilene	AP	1759	12	17	21	M
	Alice	AP	180	26	30	34	M
	Amarillo	AP	3607	2	8	12	M
	Austin	AP	597	19	25	29	M
	Big Spring	AP	2537	12	18	22	M
	Brownsville	AP	16	32	36	40	M
	Bryan	AP	275	22	27	31	M

Table 1 .... Winter Climatic Conditions (*Continued*)[a]

Col. 1 State	Col. 2 Station	Col. 3[b]	Col. 4[c] Elev, Ft.	Col. 5 Design Winter Temperatures Median of Extremes	99%	97½%	Col. 6[d] Coincident Wind Velocity
Texas (Con't)	Corpus Christi	AP	43r	28	32	36	M
	Dallas	AP	481	14	19	24	H
	Del Rio, Laughlin	AFB	1072	24	28	31	M
	El Paso	AP	3918	16	21	25	L
	Fort Worth	AP	544r	14	20	24	H
	Galveston	AP	5r	28	32	36	M
	Houston	AP	50	23	28	32	M
	Laredo	AFB	503	29	32	36	L
	Lubbock	AP	3243	4	11	15	M
	Lufkin	AP	286	19	24	28	M
	Midland	AP	2851r	13	19	23	M
	Mineral Wells	AP	934	12	18	22	H
	Port Arthur	AP	16	25	29	33	M
	San Angelo, Goodfellow	AFB	1878	15	20	25	M
	San Antonio	AP	792	22	25	30	L
	Sherman, Perrin	AFB	763	12	18	23	H
	Tyler	AP	527	12	17	21	M
	Victoria	AP	104	24	28	32	M
	Waco	AP	500	16	21	26	M
	Wichita Falls	AP	994	9	15	19	H
Utah	Cedar City	AP	5613	−10	−1	6	VL
	Delta	AP	4759	−9	0	4	VL
	Milford	AP	5028	−14	−5	−1	VL
	Ogden	CO	4400	−3	7	11	VL
	St. George	CO	2899	13	22	26	VL
	Salt Lake City	AP	4220	−2	5	9	L
Vt.	Burlington	AP	331	−18	−12	−7	M
Va.	Danville	AP	590	9	13	17	L
	Lynchburg	AP	947	10	15	19	L
	Norfolk	AP	26	18	20	23	M
	Richmond	AP	162	10	14	18	L
	Roanoke	AP	1174r	9	15	18	L
Wash.	Bellingham	AP	150	8	14	18	L
	Ellensburg	AP	1729	−5	2	6	VL
	Everett, Paine	AFB	598	13	19	24	L
	Moses Lake, Larson	AFB	1183	−14	−7	−1	VL
	Olympia	AP	190	15	21	25	L
	Seattle, Boeing Field	AP	14	17	23	27	L
	Seattle-Tacoma	AP	386	14	20	24	L
	Seattle	CO	14r	22	28	32	L
	Spokane	AP	2357	−5	−2	4	VL
	Walla Walla	AP	1185	5	12	16	VL
	Yakima	AP	1061	−1	6	10	VL
W. Va.	Charleston	AP	939	1	9	14	L
	Elkins	AP	1970	−4	1	5	L
	Huntington	CO	565r	4	10	14	L
	Martinsburg	AP	537	1	6	10	L
	Morgantown	AP	1245	−2	3	7	L
	Parkersburg	CO	615r	2	8	12	L
Wis.	Eau Claire	AP	888	−21	−15	−11	L
	Green Bay	AP	683	−16	−12	−7	M
	La Crosse	AP	652	−18	−12	−8	M
	Madison	AP	858	−13	−9	−5	M
	Milwaukee	AP	672	−11	−6	−2	M
	Wausau	AP	1196	−24	−18	−14	M
Wyo.	Casper	AP	5319	−20	−11	−5	L
	Cheyenne	AP	6126	−15	−6	−2	M
	Laramie	AP	7266	−17	−6	−2	M
	Rock Springs	AP	6741	−16	−6	−1	VL
	Sheridan	AP	3942	−21	−12	−7	L

Table 1 .... Winter Climatic Conditions (Concluded)[a]

CANADA

Col. 1 Prov.	Col. 2 Station	Col. 3[b]	Col. 4[c] Elev, Ft.	Col. 5 Average Ann. Min., F	Col. 6 January Design Temp		Col. 7[d] Coincident Wind Velocity
					99%	97½%	
Alta.	Calgary	AP	3540	−30	−29	−25	M
	Edmonton	AP	2219	−30	−29	−26	VL
	Grande Prairie	AP	2190	−44	−43	−37	VL
	Lethbridge	AP	3018	−31	−31	−24	M
	Medicine Hat	AP	2365	−33	−30	−26	M
B. C.	Fort Nelson	AP	1230	−43	−44	−41	VL
	Penticton	AP	1121	0	−1	3	L
	Prince George	AP	2218	−38	−37	−31	VL
	Prince Rupert	*	170	7	11	15	L
	Vancouver	AP	16	13	15	19	L
	Victoria	CO	228	20	20	23	M
Man.	Brandon	*	1200	−36	−29	−26	M
	Churchill	AP	115	−43	−40	−38	H
	Dauphin	AP	999	−35	−29	−26	M
	The Pas	AP	894	−41	−35	−32	M
	Winnipeg	AP	786	−31	−28	−25	M
N.B.	*Campbellton*	CO	25	−20	−18	−14	L
	Fredericton	AP	74	−19	−16	−10	L
	Moncton	AP	248	−16	−12	−7	H
	Saint John	AP	352	−15	−12	−7	M
Nfld.	Corner Brook	*	40	−15	−10	−5	H
	Gander	AP	482	−5	−5	−1	H
	Goose	AP	144	−28	−27	−25	M
	St. John's	AP	463	1	2	6	H
N.W.T.	Fort Smith	AP	665	−51	−49	−46	VL
	Frobisher Bay	AP	68	−45	−45	−42	H
	Norman Wells	AP	209	−53	−55	−52	VL
	Resolute	*	209	−53	−49	−47	M
	Yellowknife	AP	682	−51	−49	−47	VL
N.S.	Halifax	AP	136	−4	0	4	H
	Sydney	AP	197	−3	0	5	H
	Yarmouth	AP	136	2	5	9	H
Ont.	Fort William	AP	644	−31	−27	−23	L
	Hamilton	*	303	−6	0	3	M
	Kapuskasing	AP	752	−37	−31	−28	M
	Kingston	*	300	−16	−10	−7	M
	Kitchener	*	1125	−10	−3	1	M
	London	AP	912	−9	−1	3	M
	North Bay	AP	1210	−27	−21	−17	M
	Oshawa	*	370	−11	−5	−2	M
	Ottawa	AP	413	−21	−17	−13	M
	Peterborough	*	648	−20	−13	−9	M
	St. Catharine's	*	325	−4	2	5	M
	Sudbury	*	850	−25	−20	−15	VL
	Toronto	AP	578	−10	−3	1	M
	Windsor	AP	637	−1	4	7	M
P.E.I.	Charlottetown	AP	186	−11	−6	−3	H
Que.	Bagotville	*	536	−30	−24	−20	VL
	Knob Lake	AP	1681	−47	−40	−37	M
	Megantic	AP	1362	−27	−20	−16	M
	Mont Joli	AP	150	−16	−16	−12	H
	Montreal	AP	98	−20	−16	−10	M
	Quebec	AP	245	−25	−19	−13	M
	Seven Islands	AP	190	−29	−27	−22	L
	Sherbrooke	*	620	−24	−18	−13	L
	Three Rivers	*	200	−24	−18	−13	M
	Val d'Or	AP	1108	−37	−31	−27	L
Sask.	Prince Albert	AP	1414	−45	−41	−35	VL
	Regina	AP	1884	−38	−34	−29	M
	Saskatoon	AP	1645	−37	−34	−30	M
	Swift Current	AP	2677	−31	−29	−25	M
Y.T.	Whitehorse	AP	2289	−45	−45	−42	VL

Reprinted By Permission From
ASHRAE Guide and Data Book 1965

Appendix: Compound Interest Factors

1% Compound Interest Factors 1%

	Single Payment		Uniform Payment Series				Arithmetic Gradient		
	Compound Amount Factor	Present Worth Factor	Sinking Fund Factor	Capital Recovery Factor	Compound Amount Factor	Present Worth Factor	Gradient Uniform Series	Gradient Present Worth	
n	Find *F* Given *P* *F*/*P*	Find *P* Given *F* *P*/*F*	Find *A* Given *F* *A*/*F*	Find *A* Given *P* *A*/*P*	Find *F* Given *A* *F*/*A*	Find *P* Given *A* *P*/*A*	Find *A* Given *G* *A*/*G*	Find *P* Given *G* *P*/*G*	*n*
1	1.010	.9901	1.0000	1.0100	1.000	0.990	0	0	**1**
2	1.020	.9803	.4975	.5075	2.010	1.970	0.498	0.980	**2**
3	1.030	.9706	.3300	.3400	3.030	2.941	0.993	2.921	**3**
4	1.041	.9610	.2463	.2563	4.060	3.902	1.488	5.804	**4**
5	1.051	.9515	.1960	.2060	5.101	4.853	1.980	9.610	**5**
6	1.062	.9420	.1625	.1725	6.152	5.795	2.471	14.320	**6**
7	1.072	.9327	.1386	.1486	7.214	6.728	2.960	19.917	**7**
8	1.083	.9235	.1207	.1307	8.286	7.652	3.448	26.381	**8**
9	1.094	.9143	.1067	.1167	9.369	8.566	3.934	33.695	**9**
10	1.105	.9053	.0956	.1056	10.462	9.471	4.418	41.843	**10**
11	1.116	.8963	.0865	.0965	11.567	10.368	4.900	50.806	**11**
12	1.127	.8874	.0788	.0888	12.682	11.255	5.381	60.568	**12**
13	1.138	.8787	.0724	.0824	13.809	12.134	5.861	71.112	**13**
14	1.149	.8700	.0669	.0769	14.947	13.004	6.338	82.422	**14**
15	1.161	.8613	.0621	.0721	16.097	13.865	6.814	94.481	**15**
16	1.173	.8528	.0579	.0679	17.258	14.718	7.289	107.273	**16**
17	1.184	.8444	.0543	.0643	18.430	15.562	7.761	120.783	**17**
18	1.196	.8360	.0510	.0610	19.615	16.398	8.232	134.995	**18**
19	1.208	.8277	.0481	.0581	20.811	17.226	8.702	149.895	**19**
20	1.220	.8195	.0454	.0554	22.019	18.046	9.169	165.465	**20**
21	1.232	.8114	.0430	.0530	23.239	18.857	9.635	181.694	**21**
22	1.245	.8034	.0409	.0509	24.472	19.660	10.100	198.565	**22**
23	1.257	.7954	.0389	.0489	25.716	20.456	10.563	216.065	**23**
24	1.270	.7876	.0371	.0471	26.973	21.243	11.024	234.179	**24**
25	1.282	.7798	.0354	.0454	28.243	22.023	11.483	252.892	**25**
26	1.295	.7720	.0339	.0439	29.526	22.795	11.941	272.195	**26**
27	1.308	.7644	.0324	.0424	30.821	23.560	12.397	292.069	**27**
28	1.321	.7568	.0311	.0411	32.129	24.316	12.852	312.504	**28**
29	1.335	.7493	.0299	.0399	33.450	25.066	13.304	333.486	**29**
30	1.348	.7419	.0287	.0387	34.785	25.808	13.756	355.001	**30**
36	1.431	.6989	.0232	.0332	43.077	30.107	16.428	494.620	**36**
40	1.489	.6717	.0205	.0305	48.886	32.835	18.178	596.854	**40**
48	1.612	.6203	.0163	.0263	61.223	37.974	21.598	820.144	**48**
50	1.645	.6080	.0155	.0255	64.463	39.196	22.436	879.417	**50**
52	1.678	.5961	.0148	.0248	67.769	40.394	23.269	939.916	**52**
60	1.817	.5504	.0122	.0222	81.670	44.955	26.533	1 192.80	**60**
70	2.007	.4983	.00993	.0199	100.676	50.168	30.470	1 528.64	**70**
72	2.047	.4885	.00955	.0196	104.710	51.150	31.239	1 597.86	**72**
80	2.217	.4511	.00822	.0182	121.671	54.888	34.249	1 879.87	**80**
84	2.307	.4335	.00765	.0177	130.672	56.648	35.717	2 023.31	**84**
90	2.449	.4084	.00690	.0169	144.863	59.161	37.872	2 240.56	**90**
96	2.599	.3847	.00625	.0163	159.927	61.528	39.973	2 459.42	**96**
100	2.705	.3697	.00587	.0159	170.481	63.029	41.343	2 605.77	**100**
104	2.815	.3553	.00551	.0155	181.464	64.471	42.688	2 752.17	**104**
120	3.300	.3030	.00435	.0143	230.039	69.701	47.835	3 334.11	**120**
240	10.893	.0918	.00101	.0110	989.254	90.819	75.739	6 878.59	**240**
360	35.950	.0278	.00029	.0103	3 495.0	97.218	89.699	8 720.43	**360**
480	118.648	.00843	.00008	.0101	11 764.8	99.157	95.920	9 511.15	**480**

2% Compound Interest Factors 2%

n	Single Payment		Uniform Payment Series				Arithmetic Gradient		n
	Compound Amount Factor	Present Worth Factor	Sinking Fund Factor	Capital Recovery Factor	Compound Amount Factor	Present Worth Factor	Gradient Uniform Series	Gradient Present Worth	
	Find *F* Given *P* *F/P*	Find *P* Given *F* *P/F*	Find *A* Given *F* *A/F*	Find *A* Given *P* *A/P*	Find *F* Given *A* *F/A*	Find *P* Given *A* *P/A*	Find *A* Given *G* *A/G*	Find *P* Given *G* *P/G*	
1	1.020	.9804	1.0000	1.0200	1.000	0.980	0	0	**1**
2	1.040	.9612	.4951	.5151	2.020	1.942	0.495	0.961	**2**
3	1.061	.9423	.3268	.3468	3.060	2.884	0.987	2.846	**3**
4	1.082	.9238	.2426	.2626	4.122	3.808	1.475	5.617	**4**
5	1.104	.9057	.1922	.2122	5.204	4.713	1.960	9.240	**5**
6	1.126	.8880	.1585	.1785	6.308	5.601	2.442	13.679	**6**
7	1.149	.8706	.1345	.1545	7.434	6.472	2.921	18.903	**7**
8	1.172	.8535	.1165	.1365	8.583	7.325	3.396	24.877	**8**
9	1.195	.8368	.1025	.1225	9.755	8.162	3.868	31.571	**9**
10	1.219	.8203	.0913	.1113	10.950	8.983	4.337	38.954	**10**
11	1.243	.8043	.0822	.1022	12.169	9.787	4.802	46.996	**11**
12	1.268	.7885	.0746	.0946	13.412	10.575	5.264	55.669	**12**
13	1.294	.7730	.0681	.0881	14.680	11.348	5.723	64.946	**13**
14	1.319	.7579	.0626	.0826	15.974	12.106	6.178	74.798	**14**
15	1.346	.7430	.0578	.0778	17.293	12.849	6.631	85.200	**15**
16	1.373	.7284	.0537	.0737	18.639	13.578	7.080	96.127	**16**
17	1.400	.7142	.0500	.0700	20.012	14.292	7.526	107.553	**17**
18	1.428	.7002	.0467	.0667	21.412	14.992	7.968	119.456	**18**
19	1.457	.6864	.0438	.0638	22.840	15.678	8.407	131.812	**19**
20	1.486	.6730	.0412	.0612	24.297	16.351	8.843	144.598	**20**
21	1.516	.6598	.0388	.0588	25.783	17.011	9.276	157.793	**21**
22	1.546	.6468	.0366	.0566	27.299	17.658	9.705	171.377	**22**
23	1.577	.6342	.0347	.0547	28.845	18.292	10.132	185.328	**23**
24	1.608	.6217	.0329	.0529	30.422	18.914	10.555	199.628	**24**
25	1.641	.6095	.0312	.0512	32.030	19.523	10.974	214.256	**25**
26	1.673	.5976	.0297	.0497	33.671	20.121	11.391	229.196	**26**
27	1.707	.5859	.0283	.0483	35.344	20.707	11.804	244.428	**27**
28	1.741	.5744	.0270	.0470	37.051	21.281	12.214	259.936	**28**
29	1.776	.5631	.0258	.0458	38.792	21.844	12.621	275.703	**29**
30	1.811	.5521	.0247	.0447	40.568	22.396	13.025	291.713	**30**
36	2.040	.4902	.0192	.0392	51.994	25.489	15.381	392.036	**36**
40	2.208	.4529	.0166	.0366	60.402	27.355	16.888	461.989	**40**
48	2.587	.3865	.0126	.0326	79.353	30.673	19.755	605.961	**48**
50	2.692	.3715	.0118	.0318	84.579	31.424	20.442	642.355	**50**
52	2.800	.3571	.0111	.0311	90.016	32.145	21.116	678.779	**52**
60	3.281	.3048	.00877	.0288	114.051	34.761	23.696	823.692	**60**
70	4.000	.2500	.00667	.0267	149.977	37.499	26.663	999.829	**70**
72	4.161	.2403	.00633	.0263	158.056	37.984	27.223	1 034.050	**72**
80	4.875	.2051	.00516	.0252	193.771	39.744	29.357	1 166.781	**80**
84	5.277	.1895	.00468	.0247	213.865	40.525	30.361	1 230.413	**84**
90	5.943	.1683	.00405	.0240	247.155	41.587	31.793	1 322.164	**90**
96	6.693	.1494	.00351	.0235	284.645	42.529	33.137	1 409.291	**96**
100	7.245	.1380	.00320	.0232	312.230	43.098	33.986	1 464.747	**100**
104	7.842	.1275	.00292	.0229	342.090	43.624	34.799	1 518.082	**104**
120	10.765	.0929	.00205	.0220	488.255	45.355	37.711	1 710.411	**120**
240	115.887	.00863	.00017	.0202	5 744.4	49.569	47.911	2 374.878	**240**
360	1 247.5	.00080	.00002	.0200	62 326.8	49.960	49.711	2 483.567	**360**
480	13 429.8	.00007		.0200	671 442.0	49.996	49.964	2 498.027	**480**

3% Compound Interest Factors 3%

	Single Payment		Uniform Payment Series				Arithmetic Gradient		
	Compound Amount Factor	Present Worth Factor	Sinking Fund Factor	Capital Recovery Factor	Compound Amount Factor	Present Worth Factor	Gradient Uniform Series	Gradient Present Worth	
n	Find *F* Given *P* *F/P*	Find *P* Given *F* *P/F*	Find *A* Given *F* *A/F*	Find *A* Given *P* *A/P*	Find *F* Given *A* *F/A*	Find *P* Given *A* *P/A*	Find *A* Given *G* *A/G*	Find *P* Given *G* *P/G*	*n*
1	1.030	.9709	1.0000	1.0300	1.000	0.971	0	0	**1**
2	1.061	.9426	.4926	.5226	2.030	1.913	0.493	0.943	**2**
3	1.093	.9151	.3235	.3535	3.091	2.829	0.980	2.773	**3**
4	1.126	.8885	.2390	.2690	4.184	3.717	1.463	5.438	**4**
5	1.159	.8626	.1884	.2184	5.309	4.580	1.941	8.889	**5**
6	1.194	.8375	.1546	.1846	6.468	5.417	2.414	13.076	**6**
7	1.230	.8131	.1305	.1605	7.662	6.230	2.882	17.955	**7**
8	1.267	.7894	.1125	.1425	8.892	7.020	3.345	23.481	**8**
9	1.305	.7664	.0984	.1284	10.159	7.786	3.803	29.612	**9**
10	1.344	.7441	.0872	.1172	11.464	8.530	4.256	36.309	**10**
11	1.384	.7224	.0781	.1081	12.808	9.253	4.705	43.533	**11**
12	1.426	.7014	.0705	.1005	14.192	9.954	5.148	51.248	**12**
13	1.469	.6810	.0640	.0940	15.618	10.635	5.587	59.419	**13**
14	1.513	.6611	.0585	.0885	17.086	11.296	6.021	68.014	**14**
15	1.558	.6419	.0538	.0838	18.599	11.938	6.450	77.000	**15**
16	1.605	.6232	.0496	.0796	20.157	12.561	6.874	86.348	**16**
17	1.653	.6050	.0460	.0760	21.762	13.166	7.294	96.028	**17**
18	1.702	.5874	.0427	.0727	23.414	13.754	7.708	106.014	**18**
19	1.754	.5703	.0398	.0698	25.117	14.324	8.118	116.279	**19**
20	1.806	.5537	.0372	.0672	26.870	14.877	8.523	126.799	**20**
21	1.860	.5375	.0349	.0649	28.676	15.415	8.923	137.549	**21**
22	1.916	.5219	.0327	.0627	30.537	15.937	9.319	148.509	**22**
23	1.974	.5067	.0308	.0608	32.453	16.444	9.709	159.656	**23**
24	2.033	.4919	.0290	.0590	34.426	16.936	10.095	170.971	**24**
25	2.094	.4776	.0274	.0574	36.459	17.413	10.477	182.433	**25**
26	2.157	.4637	.0259	.0559	38.553	17.877	10.853	194.026	**26**
27	2.221	.4502	.0246	.0546	40.710	18.327	11.226	205.731	**27**
28	2.288	.4371	.0233	.0533	42.931	18.764	11.593	217.532	**28**
29	2.357	.4243	.0221	.0521	45.219	19.188	11.956	229.413	**29**
30	2.427	.4120	.0210	.0510	47.575	19.600	12.314	241.361	**30**
31	2.500	.4000	.0200	.0500	50.003	20.000	12.668	253.361	**31**
32	2.575	.3883	.0190	.0490	52.503	20.389	13.017	265.399	**32**
33	2.652	.3770	.0182	.0482	55.078	20.766	13.362	277.464	**33**
34	2.732	.3660	.0173	.0473	57.730	21.132	13.702	289.544	**34**
35	2.814	.3554	.0165	.0465	60.462	21.487	14.037	301.627	**35**
40	3.262	.3066	.0133	.0433	75.401	23.115	15.650	361.750	**40**
45	3.782	.2644	.0108	.0408	92.720	24.519	17.156	420.632	**45**
50	4.384	.2281	.00887	.0389	112.797	25.730	18.558	477.480	**50**
55	5.082	.1968	.00735	.0373	136.072	26.774	19.860	531.741	**55**
60	5.892	.1697	.00613	.0361	163.053	27.676	21.067	583.052	**60**
65	6.830	.1464	.00515	.0351	194.333	28.453	22.184	631.201	**65**
70	7.918	.1263	.00434	.0343	230.594	29.123	23.215	676.087	**70**
75	9.179	.1089	.00367	.0337	272.631	29.702	24.163	717.698	**75**
80	10.641	.0940	.00311	.0331	321.363	30.201	25.035	756.086	**80**
85	12.336	.0811	.00265	.0326	377.857	30.631	25.835	791.353	**85**
90	14.300	.0699	.00226	.0323	443.349	31.002	26.567	823.630	**90**
95	16.578	.0603	.00193	.0319	519.272	31.323	27.235	853.074	**95**
100	19.219	.0520	.00165	.0316	607.287	31.599	27.844	879.854	**100**

3½% Compound Interest Factors 3½%

n	Single Payment		Uniform Payment Series				Arithmetic Gradient		n
	Compound Amount Factor	Present Worth Factor	Sinking Fund Factor	Capital Recovery Factor	Compound Amount Factor	Present Worth Factor	Gradient Uniform Series	Gradient Present Worth	
	Find *F* Given *P* *F/P*	Find *P* Given *F* *P/F*	Find *A* Given *F* *A/F*	Find *A* Given *P* *A/P*	Find *F* Given *A* *F/A*	Find *P* Given *A* *P/A*	Find *A* Given *G* *A/G*	Find *P* Given *G* *P/G*	n
1	1.035	.9662	1.0000	1.0350	1.000	0.966	0	0	**1**
2	1.071	.9335	.4914	.5264	2.035	1.900	0.491	0.933	**2**
3	1.109	.9019	.3219	.3569	3.106	2.802	0.977	2.737	**3**
4	1.148	.8714	.2373	.2723	4.215	3.673	1.457	5.352	**4**
5	1.188	.8420	.1865	.2215	5.362	4.515	1.931	8.719	**5**
6	1.229	.8135	.1527	.1877	6.550	5.329	2.400	12.787	**6**
7	1.272	.7860	.1285	.1635	7.779	6.115	2.862	17.503	**7**
8	1.317	.7594	.1105	.1455	9.052	6.874	3.320	22.819	**8**
9	1.363	.7337	.0964	.1314	10.368	7.608	3.771	28.688	**9**
10	1.411	.7089	.0852	.1202	11.731	8.317	4.217	35.069	**10**
11	1.460	.6849	.0761	.1111	13.142	9.002	4.657	41.918	**11**
12	1.511	.6618	.0685	.1035	14.602	9.663	5.091	49.198	**12**
13	1.564	.6394	.0621	.0971	16.113	10.303	5.520	56.871	**13**
14	1.619	.6178	.0566	.0916	17.677	10.921	5.943	64.902	**14**
15	1.675	.5969	.0518	.0868	19.296	11.517	6.361	73.258	**15**
16	1.734	.5767	.0477	.0827	20.971	12.094	6.773	81.909	**16**
17	1.795	.5572	.0440	.0790	22.705	12.651	7.179	90.824	**17**
18	1.857	.5384	.0408	.0758	24.500	13.190	7.580	99.976	**18**
19	1.922	.5202	.0379	.0729	26.357	13.710	7.975	109.339	**19**
20	1.990	.5026	.0354	.0704	28.280	14.212	8.365	118.888	**20**
21	2.059	.4856	.0330	.0680	30.269	14.698	8.749	128.599	**21**
22	2.132	.4692	.0309	.0659	32.329	15.167	9.128	138.451	**22**
23	2.206	.4533	.0290	.0640	34.460	15.620	9.502	148.423	**23**
24	2.283	.4380	.0273	.0623	36.666	16.058	9.870	158.496	**24**
25	2.363	.4231	.0257	.0607	38.950	16.482	10.233	168.652	**25**
26	2.446	.4088	.0242	.0592	41.313	16.890	10.590	178.873	**26**
27	2.532	.3950	.0229	.0579	43.759	17.285	10.942	189.143	**27**
28	2.620	.3817	.0216	.0566	46.291	17.667	11.289	199.448	**28**
29	2.712	.3687	.0204	.0554	48.911	18.036	11.631	209.773	**29**
30	2.807	.3563	.0194	.0544	51.623	18.392	11.967	220.105	**30**
31	2.905	.3442	.0184	.0534	54.429	18.736	12.299	230.432	**31**
32	3.007	.3326	.0174	.0524	57.334	19.069	12.625	240.742	**32**
33	3.112	.3213	.0166	.0516	60.341	19.390	12.946	251.025	**33**
34	3.221	.3105	.0158	.0508	63.453	19.701	13.262	261.271	**34**
35	3.334	.3000	.0150	.0500	66.674	20.001	13.573	271.470	**35**
40	3.959	.2526	.0118	.0468	84.550	21.355	15.055	321.490	**40**
45	4.702	.2127	.00945	.0445	105.781	22.495	16.417	369.307	**45**
50	5.585	.1791	.00763	.0426	130.998	23.456	17.666	414.369	**50**
55	6.633	.1508	.00621	.0412	160.946	24.264	18.808	456.352	**55**
60	7.878	.1269	.00509	.0401	196.516	24.945	19.848	495.104	**60**
65	9.357	.1069	.00419	.0392	238.762	25.518	20.793	530.598	**65**
70	11.113	.0900	.00346	.0385	288.937	26.000	21.650	562.895	**70**
75	13.199	.0758	.00287	.0379	348.529	26.407	22.423	592.121	**75**
80	15.676	.0638	.00238	.0374	419.305	26.749	23.120	618.438	**80**
85	18.618	.0537	.00199	.0370	503.365	27.037	23.747	642.036	**85**
90	22.112	.0452	.00166	.0367	603.202	27.279	24.308	663.118	**90**
95	26.262	.0381	.00139	.0364	721.778	27.483	24.811	681.890	**95**
100	31.191	.0321	.00116	.0362	862.608	27.655	25.259	698.554	**100**

4% Compound Interest Factors 4%

	Single Payment		Uniform Payment Series				Arithmetic Gradient		
	Compound Amount Factor	Present Worth Factor	Sinking Fund Factor	Capital Recovery Factor	Compound Amount Factor	Present Worth Factor	Gradient Uniform Series	Gradient Present Worth	
n	Find *F* Given *P* *F/P*	Find *P* Given *F* *P/F*	Find *A* Given *F* *A/F*	Find *A* Given *P* *A/P*	Find *F* Given *A* *F/A*	Find *P* Given *A* *P/A*	Find *A* Given *G* *A/G*	Find *P* Given *G* *P/G*	*n*
1	1.040	.9615	1.0000	1.0400	1.000	0.962	0	0	**1**
2	1.082	.9246	.4902	.5302	2.040	1.886	0.490	0.925	**2**
3	1.125	.8890	.3203	.3603	3.122	2.775	0.974	2.702	**3**
4	1.170	.8548	.2355	.2755	4.246	3.630	1.451	5.267	**4**
5	1.217	.8219	.1846	.2246	5.416	4.452	1.922	8.555	**5**
6	1.265	.7903	.1508	.1908	6.633	5.242	2.386	12.506	**6**
7	1.316	.7599	.1266	.1666	7.898	6.002	2.843	17.066	**7**
8	1.369	.7307	.1085	.1485	9.214	6.733	3.294	22.180	**8**
9	1.423	.7026	.0945	.1345	10.583	7.435	3.739	27.801	**9**
10	1.480	.6756	.0833	.1233	12.006	8.111	4.177	33.881	**10**
11	1.539	.6496	.0741	.1141	13.486	8.760	4.609	40.377	**11**
12	1.601	.6246	.0666	.1066	15.026	9.385	5.034	47.248	**12**
13	1.665	.6006	.0601	.1001	16.627	9.986	5.453	54.454	**13**
14	1.732	.5775	.0547	.0947	18.292	10.563	5.866	61.962	**14**
15	1.801	.5553	.0499	.0899	20.024	11.118	6.272	69.735	**15**
16	1.873	.5339	.0458	.0858	21.825	11.652	6.672	77.744	**16**
17	1.948	.5134	.0422	.0822	23.697	12.166	7.066	85.958	**17**
18	2.026	.4936	.0390	.0790	25.645	12.659	7.453	94.350	**18**
19	2.107	.4746	.0361	.0761	27.671	13.134	7.834	102.893	**19**
20	2.191	.4564	.0336	.0736	29.778	13.590	8.209	111.564	**20**
21	2.279	.4388	.0313	.0713	31.969	14.029	8.578	120.341	**21**
22	2.370	.4220	.0292	.0692	34.248	14.451	8.941	129.202	**22**
23	2.465	.4057	.0273	.0673	36.618	14.857	9.297	138.128	**23**
24	2.563	.3901	.0256	.0656	39.083	15.247	9.648	147.101	**24**
25	2.666	.3751	.0240	.0640	41.646	15.622	9.993	156.104	**25**
26	2.772	.3607	.0226	.0626	44.312	15.983	10.331	165.121	**26**
27	2.883	.3468	.0212	.0612	47.084	16.330	10.664	174.138	**27**
28	2.999	.3335	.0200	.0600	49.968	16.663	10.991	183.142	**28**
29	3.119	.3207	.0189	.0589	52.966	16.984	11.312	192.120	**29**
30	3.243	.3083	.0178	.0578	56.085	17.292	11.627	201.062	**30**
31	3.373	.2965	.0169	.0569	59.328	17.588	11.937	209.955	**31**
32	3.508	.2851	.0159	.0559	62.701	17.874	12.241	218.792	**32**
33	3.648	.2741	.0151	.0551	66.209	18.148	12.540	227.563	**33**
34	3.794	.2636	.0143	.0543	69.858	18.411	12.832	236.260	**34**
35	3.946	.2534	.0136	.0536	73.652	18.665	13.120	244.876	**35**
40	4.801	.2083	.0105	.0505	95.025	19.793	14.476	286.530	**40**
45	5.841	.1712	.00826	.0483	121.029	20.720	15.705	325.402	**45**
50	7.107	.1407	.00655	.0466	152.667	21.482	16.812	361.163	**50**
55	8.646	.1157	.00523	.0452	191.159	22.109	17.807	393.689	**55**
60	10.520	.0951	.00420	.0442	237.990	22.623	18.697	422.996	**60**
65	12.799	.0781	.00339	.0434	294.968	23.047	19.491	449.201	**65**
70	15.572	.0642	.00275	.0427	364.290	23.395	20.196	472.479	**70**
75	18.945	.0528	.00223	.0422	448.630	23.680	20.821	493.041	**75**
80	23.050	.0434	.00181	.0418	551.244	23.915	21.372	511.116	**80**
85	28.044	.0357	.00148	.0415	676.089	24.109	21.857	526.938	**85**
90	34.119	.0293	.00121	.0412	827.981	24.267	22.283	540.737	**90**
95	41.511	.0241	.00099	.0410	1 012.8	24.398	22.655	552.730	**95**
100	50.505	.0198	.00081	.0408	1 237.6	24.505	22.980	563.125	**100**

4½% Compound Interest Factors 4½%

	Single Payment		Uniform Payment Series				Arithmetic Gradient		
	Compound Amount Factor	Present Worth Factor	Sinking Fund Factor	Capital Recovery Factor	Compound Amount Factor	Present Worth Factor	Gradient Uniform Series	Gradient Present Worth	
n	Find F Given P F/P	Find P Given F P/F	Find A Given F A/F	Find A Given P A/P	Find F Given A F/A	Find P Given A P/A	Find A Given G A/G	Find P Given G P/G	n
1	1.045	.9569	1.0000	1.0450	1.000	0.957	0	0	1
2	1.092	.9157	.4890	.5340	2.045	1.873	0.489	0.916	2
3	1.141	.8763	.3188	.3638	3.137	2.749	0.971	2.668	3
4	1.193	.8386	.2337	.2787	4.278	3.588	1.445	5.184	4
5	1.246	.8025	.1828	.2278	5.471	4.390	1.912	8.394	5
6	1.302	.7679	.1489	.1939	6.717	5.158	2.372	12.233	6
7	1.361	.7348	.1247	.1697	8.019	5.893	2.824	16.642	7
8	1.422	.7032	.1066	.1516	9.380	6.596	3.269	21.564	8
9	1.486	.6729	.0926	.1376	10.802	7.269	3.707	26.948	9
10	1.553	.6439	.0814	.1264	12.288	7.913	4.138	32.743	10
11	1.623	.6162	.0722	.1172	13.841	8.529	4.562	38.905	11
12	1.696	.5897	.0647	.1097	15.464	9.119	4.978	45.391	12
13	1.772	.5643	.0583	.1033	17.160	9.683	5.387	52.163	13
14	1.852	.5400	.0528	.0978	18.932	10.223	5.789	59.182	14
15	1.935	.5167	.0481	.0931	20.784	10.740	6.184	66.416	15
16	2.022	.4945	.0440	.0890	22.719	11.234	6.572	73.833	16
17	2.113	.4732	.0404	.0854	24.742	11.707	6.953	81.404	17
18	2.208	.4528	.0372	.0822	26.855	12.160	7.327	89.102	18
19	2.308	.4333	.0344	.0794	29.064	12.593	7.695	96.901	19
20	2.412	.4146	.0319	.0769	31.371	13.008	8.055	104.779	20
21	2.520	.3968	.0296	.0746	33.783	13.405	8.409	112.715	21
22	2.634	.3797	.0275	.0725	36.303	13.784	8.755	120.689	22
23	2.752	.3634	.0257	.0707	38.937	14.148	9.096	128.682	23
24	2.876	.3477	.0240	.0690	41.689	14.495	9.429	136.680	24
25	3.005	.3327	.0224	.0674	44.565	14.828	9.756	144.665	25
26	3.141	.3184	.0210	.0660	47.571	15.147	10.077	152.625	26
27	3.282	.3047	.0197	.0647	50.711	15.451	10.391	160.547	27
28	3.430	.2916	.0185	.0635	53.993	15.743	10.698	168.420	28
29	3.584	.2790	.0174	.0624	57.423	16.022	10.999	176.232	29
30	3.745	.2670	.0164	.0614	61.007	16.289	11.295	183.975	30
31	3.914	.2555	.0154	.0604	64.752	16.544	11.583	191.640	31
32	4.090	.2445	.0146	.0596	68.666	16.789	11.866	199.220	32
33	4.274	.2340	.0137	.0587	72.756	17.023	12.143	206.707	33
34	4.466	.2239	.0130	.0580	77.030	17.247	12.414	214.095	34
35	4.667	.2143	.0123	.0573	81.497	17.461	12.679	221.380	35
40	5.816	.1719	.00934	.0543	107.030	18.402	13.917	256.098	40
45	7.248	.1380	.00720	.0522	138.850	19.156	15.020	287.732	45
50	9.033	.1107	.00560	.0506	178.503	19.762	15.998	316.145	50
55	11.256	.0888	.00439	.0494	227.918	20.248	16.860	341.375	55
60	14.027	.0713	.00345	.0485	289.497	20.638	17.617	363.571	60
65	17.481	.0572	.00273	.0477	366.237	20.951	18.278	382.946	65
70	21.784	.0459	.00217	.0472	461.869	21.202	18.854	399.750	70
75	27.147	.0368	.00172	.0467	581.043	21.404	19.354	414.242	75
80	33.830	.0296	.00137	.0464	729.556	21.565	19.785	426.680	80
85	42.158	.0237	.00109	.0461	914.630	21.695	20.157	437.309	85
90	52.537	.0190	.00087	.0459	1 145.3	21.799	20.476	446.359	90
95	65.471	.0153	.00070	.0457	1 432.7	21.883	20.749	454.039	95
100	81.588	.0123	.00056	.0456	1 790.9	21.950	20.981	460.537	100

5% Compound Interest Factors 5%

	Single Payment		Uniform Payment Series				Arithmetic Gradient		
	Compound Amount Factor	Present Worth Factor	Sinking Fund Factor	Capital Recovery Factor	Compound Amount Factor	Present Worth Factor	Gradient Uniform Series	Gradient Present Worth	
n	Find F Given P F/P	Find P Given F P/F	Find A Given F A/F	Find A Given P A/P	Find F Given A F/A	Find P Given A P/A	Find A Given G A/G	Find P Given G P/G	n
1	1.050	.9524	1.0000	1.0500	1.000	0.952	0	0	**1**
2	1.102	.9070	.4878	.5378	2.050	1.859	0.488	0.907	**2**
3	1.158	.8638	.3172	.3672	3.152	2.723	0.967	2.635	**3**
4	1.216	.8227	.2320	.2820	4.310	3.546	1.439	5.103	**4**
5	1.276	.7835	.1810	.2310	5.526	4.329	1.902	8.237	**5**
6	1.340	.7462	.1470	.1970	6.802	5.076	2.358	11.968	**6**
7	1.407	.7107	.1228	.1728	8.142	5.786	2.805	16.232	**7**
8	1.477	.6768	.1047	.1547	9.549	6.463	3.244	20.970	**8**
9	1.551	.6446	.0907	.1407	11.027	7.108	3.676	26.127	**9**
10	1.629	.6139	.0795	.1295	12.578	7.722	4.099	31.652	**10**
11	1.710	.5847	.0704	.1204	14.207	8.306	4.514	37.499	**11**
12	1.796	.5568	.0628	.1128	15.917	8.863	4.922	43.624	**12**
13	1.886	.5303	.0565	.1065	17.713	9.394	5.321	49.988	**13**
14	1.980	.5051	.0510	.1010	19.599	9.899	5.713	56.553	**14**
15	2.079	.4810	.0463	.0963	21.579	10.380	6.097	63.288	**15**
16	2.183	.4581	.0423	.0923	23.657	10.838	6.474	70.159	**16**
17	2.292	.4363	.0387	.0887	25.840	11.274	6.842	77.140	**17**
18	2.407	.4155	.0355	.0855	28.132	11.690	7.203	84.204	**18**
19	2.527	.3957	.0327	.0827	30.539	12.085	7.557	91.327	**19**
20	2.653	.3769	.0302	.0802	33.066	12.462	7.903	98.488	**20**
21	2.786	.3589	.0280	.0780	35.719	12.821	8.242	105.667	**21**
22	2.925	.3419	.0260	.0760	38.505	13.163	8.573	112.846	**22**
23	3.072	.3256	.0241	.0741	41.430	13.489	8.897	120.008	**23**
24	3.225	.3101	.0225	.0725	44.502	13.799	9.214	127.140	**24**
25	3.386	.2953	.0210	.0710	47.727	14.094	9.524	134.227	**25**
26	3.556	.2812	.0196	.0696	51.113	14.375	9.827	141.258	**26**
27	3.733	.2678	.0183	.0683	54.669	14.643	10.122	148.222	**27**
28	3.920	.2551	.0171	.0671	58.402	14.898	10.411	155.110	**28**
29	4.116	.2429	.0160	.0660	62.323	15.141	10.694	161.912	**29**
30	4.322	.2314	.0151	.0651	66.439	15.372	10.969	168.622	**30**
31	4.538	.2204	.0141	.0641	70.761	15.593	11.238	175.233	**31**
32	4.765	.2099	.0133	.0633	75.299	15.803	11.501	181.739	**32**
33	5.003	.1999	.0125	.0625	80.063	16.003	11.757	188.135	**33**
34	5.253	.1904	.0118	.0618	85.067	16.193	12.006	194.416	**34**
35	5.516	.1813	.0111	.0611	90.320	16.374	12.250	200.580	**35**
40	7.040	.1420	.00828	.0583	120.799	17.159	13.377	229.545	**40**
45	8.985	.1113	.00626	.0563	159.699	17.774	14.364	255.314	**45**
50	11.467	.0872	.00478	.0548	209.347	18.256	15.223	277.914	**50**
55	14.636	.0683	.00367	.0537	272.711	18.633	15.966	297.510	**55**
60	18.679	.0535	.00283	.0528	353.582	18.929	16.606	314.343	**60**
65	23.840	.0419	.00219	.0522	456.795	19.161	17.154	328.691	**65**
70	30.426	.0329	.00170	.0517	588.525	19.343	17.621	340.841	**70**
75	38.832	.0258	.00132	.0513	756.649	19.485	18.018	351.072	**75**
80	49.561	.0202	.00103	.0510	971.222	19.596	18.353	359.646	**80**
85	63.254	.0158	.00080	.0508	1 245.1	19.684	18.635	366.800	**85**
90	80.730	.0124	.00063	.0506	1 594.6	19.752	18.871	372.749	**90**
95	103.034	.00971	.00049	.0505	2 040.7	19.806	19.069	377.677	**95**
100	131.500	.00760	.00038	.0504	2 610.0	19.848	19.234	381.749	**100**

6% Compound Interest Factors 6%

	Single Payment		Uniform Payment Series				Arithmetic Gradient		
	Compound Amount Factor	Present Worth Factor	Sinking Fund Factor	Capital Recovery Factor	Compound Amount Factor	Present Worth Factor	Gradient Uniform Series	Gradient Present Worth	
n	Find *F* Given *P* *F/P*	Find *P* Given *F* *P/F*	Find *A* Given *F* *A/F*	Find *A* Given *P* *A/P*	Find *F* Given *A* *F/A*	Find *P* Given *A* *P/A*	Find *A* Given *G* *A/G*	Find *P* Given *G* *P/G*	*n*
1	1.060	.9434	1.0000	1.0600	1.000	0.943	0	0	**1**
2	1.124	.8900	.4854	.5454	2.060	1.833	0.485	0.890	**2**
3	1.191	.8396	.3141	.3741	3.184	2.673	0.961	2.569	**3**
4	1.262	.7921	.2286	.2886	4.375	3.465	1.427	4.945	**4**
5	1.338	.7473	.1774	.2374	5.637	4.212	1.884	7.934	**5**
6	1.419	.7050	.1434	.2034	6.975	4.917	2.330	11.459	**6**
7	1.504	.6651	.1191	.1791	8.394	5.582	2.768	15.450	**7**
8	1.594	.6274	.1010	.1610	9.897	6.210	3.195	19.841	**8**
9	1.689	.5919	.0870	.1470	11.491	6.802	3.613	24.577	**9**
10	1.791	.5584	.0759	.1359	13.181	7.360	4.022	29.602	**10**
11	1.898	.5268	.0668	.1268	14.972	7.887	4.421	34.870	**11**
12	2.012	.4970	.0593	.1193	16.870	8.384	4.811	40.337	**12**
13	2.133	.4688	.0530	.1130	18.882	8.853	5.192	45.963	**13**
14	2.261	.4423	.0476	.1076	21.015	9.295	5.564	51.713	**14**
15	2.397	.4173	.0430	.1030	23.276	9.712	5.926	57.554	**15**
16	2.540	.3936	.0390	.0990	25.672	10.106	6.279	63.459	**16**
17	2.693	.3714	.0354	.0954	28.213	10.477	6.624	69.401	**17**
18	2.854	.3503	.0324	.0924	30.906	10.828	6.960	75.357	**18**
19	3.026	.3305	.0296	.0896	33.760	11.158	7.287	81.306	**19**
20	3.207	.3118	.0272	.0872	36.786	11.470	7.605	87.230	**20**
21	3.400	.2942	.0250	.0850	39.993	11.764	7.915	93.113	**21**
22	3.604	.2775	.0230	.0830	43.392	12.042	8.217	98.941	**22**
23	3.820	.2618	.0213	.0813	46.996	12.303	8.510	104.700	**23**
24	4.049	.2470	.0197	.0797	50.815	12.550	8.795	110.381	**24**
25	4.292	.2330	.0182	.0782	54.864	12.783	9.072	115.973	**25**
26	4.549	.2198	.0169	.0769	59.156	13.003	9.341	121.468	**26**
27	4.822	.2074	.0157	.0757	63.706	13.211	9.603	126.860	**27**
28	5.112	.1956	.0146	.0746	68.528	13.406	9.857	132.142	**28**
29	5.418	.1846	.0136	.0736	73.640	13.591	10.103	137.309	**29**
30	5.743	.1741	.0126	.0726	79.058	13.765	10.342	142.359	**30**
31	6.088	.1643	.0118	.0718	84.801	13.929	10.574	147.286	**31**
32	6.453	.1550	.0110	.0710	90.890	14.084	10.799	152.090	**32**
33	6.841	.1462	.0103	.0703	97.343	14.230	11.017	156.768	**33**
34	7.251	.1379	.00960	.0696	104.184	14.368	11.228	161.319	**34**
35	7.686	.1301	.00897	.0690	111.435	14.498	11.432	165.743	**35**
40	10.286	.0972	.00646	.0665	154.762	15.046	12.359	185.957	**40**
45	13.765	.0727	.00470	.0647	212.743	15.456	13.141	203.109	**45**
50	18.420	.0543	.00344	.0634	290.335	15.762	13.796	217.457	**50**
55	24.650	.0406	.00254	.0625	394.171	15.991	14.341	229.322	**55**
60	32.988	.0303	.00188	.0619	533.126	16.161	14.791	239.043	**60**
65	44.145	.0227	.00139	.0614	719.080	16.289	15.160	246.945	**65**
70	59.076	.0169	.00103	.0610	967.928	16.385	15.461	253.327	**70**
75	79.057	.0126	.00077	.0608	1 300.9	16.456	15.706	258.453	**75**
80	105.796	.00945	.00057	.0606	1 746.6	16.509	15.903	262.549	**80**
85	141.578	.00706	.00043	.0604	2 343.0	16.549	16.062	265.810	**85**
90	189.464	.00528	.00032	.0603	3 141.1	16.579	16.189	268.395	**90**
95	253.545	.00394	.00024	.0602	4 209.1	16.601	16.290	270.437	**95**
100	339.300	.00295	.00018	.0602	5 638.3	16.618	16.371	272.047	**100**

10% Compound Interest Factors 10%

	Single Payment		Uniform Payment Series				Arithmetic Gradient		
	Compound Amount Factor	Present Worth Factor	Sinking Fund Factor	Capital Recovery Factor	Compound Amount Factor	Present Worth Factor	Gradient Uniform Series	Gradient Present Worth	
n	Find *F* Given *P* *F/P*	Find *P* Given *F* *P/F*	Find *A* Given *F* *A/F*	Find *A* Given *P* *A/P*	Find *F* Given *A* *F/A*	Find *P* Given *A* *P/A*	Find *A* Given *G* *A/G*	Find *P* Given *G* *P/G*	*n*
1	1.100	.9091	1.0000	1.1000	1.000	0.909	0	0	**1**
2	1.210	.8264	.4762	.5762	2.100	1.736	0.476	0.826	**2**
3	1.331	.7513	.3021	.4021	3.310	2.487	0.937	2.329	**3**
4	1.464	.6830	.2155	.3155	4.641	3.170	1.381	4.378	**4**
5	1.611	.6209	.1638	.2638	6.105	3.791	1.810	6.862	**5**
6	1.772	.5645	.1296	.2296	7.716	4.355	2.224	9.684	**6**
7	1.949	.5132	.1054	.2054	9.487	4.868	2.622	12.763	**7**
8	2.144	.4665	.0874	.1874	11.436	5.335	3.004	16.029	**8**
9	2.358	.4241	.0736	.1736	13.579	5.759	3.372	19.421	**9**
10	2.594	.3855	.0627	.1627	15.937	6.145	3.725	22.891	**10**
11	2.853	.3505	.0540	.1540	18.531	6.495	4.064	26.396	**11**
12	3.138	.3186	.0468	.1468	21.384	6.814	4.388	29.901	**12**
13	3.452	.2897	.0408	.1408	24.523	7.103	4.699	33.377	**13**
14	3.797	.2633	.0357	.1357	27.975	7.367	4.996	36.801	**14**
15	4.177	.2394	.0315	.1315	31.772	7.606	5.279	40.152	**15**
16	4.595	.2176	.0278	.1278	35.950	7.824	5.549	43.416	**16**
17	5.054	.1978	.0247	.1247	40.545	8.022	5.807	46.582	**17**
18	5.560	.1799	.0219	.1219	45.599	8.201	6.053	49.640	**18**
19	6.116	.1635	.0195	.1195	51.159	8.365	6.286	52.583	**19**
20	6.728	.1486	.0175	.1175	57.275	8.514	6.508	55.407	**20**
21	7.400	.1351	.0156	.1156	64.003	8.649	6.719	58.110	**21**
22	8.140	.1228	.0140	.1140	71.403	8.772	6.919	60.689	**22**
23	8.954	.1117	.0126	.1126	79.543	8.883	7.108	63.146	**23**
24	9.850	.1015	.0113	.1113	88.497	8.985	7.288	65.481	**24**
25	10.835	.0923	.0102	.1102	98.347	9.077	7.458	67.696	**25**
26	11.918	.0839	.00916	.1092	109.182	9.161	7.619	69.794	**26**
27	13.110	.0763	.00826	.1083	121.100	9.237	7.770	71.777	**27**
28	14.421	.0693	.00745	.1075	134.210	9.307	7.914	73.650	**28**
29	15.863	.0630	.00673	.1067	148.631	9.370	8.049	75.415	**29**
30	17.449	.0573	.00608	.1061	164.494	9.427	8.176	77.077	**30**
31	19.194	.0521	.00550	.1055	181.944	9.479	8.296	78.640	**31**
32	21.114	.0474	.00497	.1050	201.138	9.526	8.409	80.108	**32**
33	23.225	.0431	.00450	.1045	222.252	9.569	8.515	81.486	**33**
34	25.548	.0391	.00407	.1041	245.477	9.609	8.615	82.777	**34**
35	28.102	.0356	.00369	.1037	271.025	9.644	8.709	83.987	**35**
40	45.259	.0221	.00226	.1023	442.593	9.779	9.096	88.953	**40**
45	72.891	.0137	.00139	.1014	718.905	9.863	9.374	92.454	**45**
50	117.391	.00852	.00086	.1009	1 163.9	9.915	9.570	94.889	**50**
55	189.059	.00529	.00053	.1005	1 880.6	9.947	9.708	96.562	**55**
60	304.482	.00328	.00033	.1003	3 034.8	9.967	9.802	97.701	**60**
65	490.371	.00204	.00020	.1002	4 893.7	9.980	9.867	98.471	**65**
70	789.748	.00127	.00013	.1001	7 887.5	9.987	9.911	98.987	**70**
75	1 271.9	.00079	.00008	.1001	12 709.0	9.992	9.941	99.332	**75**
80	2 048.4	.00049	.00005	.1000	20 474.0	9.995	9.961	99.561	**80**
85	3 299.0	.00030	.00003	.1000	32 979.7	9.997	9.974	99.712	**85**
90	5 313.0	.00019	.00002	.1000	53 120.3	9.998	9.983	99.812	**90**
95	8 556.7	.00012	.00001	.1000	85 556.9	9.999	9.989	99.877	**95**
100	13 780.6	.00007	.00001	.1000	137 796.3	9.999	9.993	99.920	**100**

15% Compound Interest Factors 15%

	Single Payment		Uniform Payment Series				Arithmetic Gradient		
	Compound Amount Factor	Present Worth Factor	Sinking Fund Factor	Capital Recovery Factor	Compound Amount Factor	Present Worth Factor	Gradient Uniform Series	Gradient Present Worth	
n	Find *F* Given *P* *F/P*	Find *P* Given *F* *P/F*	Find *A* Given *F* *A/F*	Find *A* Given *P* *A/P*	Find *F* Given *A* *F/A*	Find *P* Given *A* *P/A*	Find *A* Given *G* *A/G*	Find *P* Given *G* *P/G*	*n*
1	1.150	.8696	1.0000	1.1500	1.000	0.870	0	0	**1**
2	1.322	.7561	.4651	.6151	2.150	1.626	0.465	0.756	**2**
3	1.521	.6575	.2880	.4380	3.472	2.283	0.907	2.071	**3**
4	1.749	.5718	.2003	.3503	4.993	2.855	1.326	3.786	**4**
5	2.011	.4972	.1483	.2983	6.742	3.352	1.723	5.775	**5**
6	2.313	.4323	.1142	.2642	8.754	3.784	2.097	7.937	**6**
7	2.660	.3759	.0904	.2404	11.067	4.160	2.450	10.192	**7**
8	3.059	.3269	.0729	.2229	13.727	4.487	2.781	12.481	**8**
9	3.518	.2843	.0596	.2096	16.786	4.772	3.092	14.755	**9**
10	4.046	.2472	.0493	.1993	20.304	5.019	3.383	16.979	**10**
11	4.652	.2149	.0411	.1911	24.349	5.234	3.655	19.129	**11**
12	5.350	.1869	.0345	.1845	29.002	5.421	3.908	21.185	**12**
13	6.153	.1625	.0291	.1791	34.352	5.583	4.144	23.135	**13**
14	7.076	.1413	.0247	.1747	40.505	5.724	4.362	24.972	**14**
15	8.137	.1229	.0210	.1710	47.580	5.847	4.565	26.693	**15**
16	9.358	.1069	.0179	.1679	55.717	5.954	4.752	28.296	**16**
17	10.761	.0929	.0154	.1654	65.075	6.047	4.925	29.783	**17**
18	12.375	.0808	.0132	.1632	75.836	6.128	5.084	31.156	**18**
19	14.232	.0703	.0113	.1613	88.212	6.198	5.231	32.421	**19**
20	16.367	.0611	.00976	.1598	102.444	6.259	5.365	33.582	**20**
21	18.822	.0531	.00842	.1584	118.810	6.312	5.488	34.645	**21**
22	21.645	.0462	.00727	.1573	137.632	6.359	5.601	35.615	**22**
23	24.891	.0402	.00628	.1563	159.276	6.399	5.704	36.499	**23**
24	28.625	.0349	.00543	.1554	184.168	6.434	5.798	37.302	**24**
25	32.919	.0304	.00470	.1547	212.793	6.464	5.883	38.031	**25**
26	37.857	.0264	.00407	.1541	245.712	6.491	5.961	38.692	**26**
27	43.535	.0230	.00353	.1535	283.569	6.514	6.032	39.289	**27**
28	50.066	.0200	.00306	.1531	327.104	6.534	6.096	39.828	**28**
29	57.575	.0174	.00265	.1527	377.170	6.551	6.154	40.315	**29**
30	66.212	.0151	.00230	.1523	434.745	6.566	6.207	40.753	**30**
31	76.144	.0131	.00200	.1520	500.957	6.579	6.254	41.147	**31**
32	87.565	.0114	.00173	.1517	577.100	6.591	6.297	41.501	**32**
33	100.700	.00993	.00150	.1515	664.666	6.600	6.336	41.818	**33**
34	115.805	.00864	.00131	.1513	765.365	6.609	6.371	42.103	**34**
35	133.176	.00751	.00113	.1511	881.170	6.617	6.402	42.359	**35**
40	267.864	.00373	.00056	.1506	1 779.1	6.642	6.517	43.283	**40**
45	538.769	.00186	.00028	.1503	3 585.1	6.654	6.583	43.805	**45**
50	1 083.7	.00092	.00014	.1501	7 217.7	6.661	6.620	44.096	**50**
55	2 179.6	.00046	.00007	.1501	14 524.1	6.664	6.641	44.256	**55**
60	4 384.0	.00023	.00003	.1500	29 220.0	6.665	6.653	44.343	**60**
65	8 817.8	.00011	.00002	.1500	58 778.6	6.666	6.659	44.390	**65**
70	17 735.7	.00006	.00001	.1500	118 231.5	6.666	6.663	44.416	**70**
75	35 672.9	.00003		.1500	237 812.5	6.666	6.665	44.429	**75**
80	71 750.9	.00001		.1500	478 332.6	6.667	6.666	44.436	**80**
85	144 316.7	.00001		.1500	962 104.4	6.667	6.666	44.440	**85**

20% Compound Interest Factors 20%

	Single Payment		Uniform Payment Series				Arithmetic Gradient		
	Compound Amount Factor	Present Worth Factor	Sinking Fund Factor	Capital Recovery Factor	Compound Amount Factor	Present Worth Factor	Gradient Uniform Series	Gradient Present Worth	
n	Find *F* Given *P* *F*/*P*	Find *P* Given *F* *P*/*F*	Find *A* Given *F* *A*/*F*	Find *A* Given *P* *A*/*P*	Find *F* Given *A* *F*/*A*	Find *P* Given *A* *P*/*A*	Find *A* Given *G* *A*/*G*	Find *P* Given *G* *P*/*G*	*n*
1	1.200	.8333	1.0000	1.2000	1.000	0.833	0	0	**1**
2	1.440	.6944	.4545	.6545	2.200	1.528	0.455	0.694	**2**
3	1.728	.5787	.2747	.4747	3.640	2.106	0.879	1.852	**3**
4	2.074	.4823	.1863	.3863	5.368	2.589	1.274	3.299	**4**
5	2.488	.4019	.1344	.3344	7.442	2.991	1.641	4.906	**5**
6	2.986	.3349	.1007	.3007	9.930	3.326	1.979	6.581	**6**
7	3.583	.2791	.0774	.2774	12.916	3.605	2.290	8.255	**7**
8	4.300	.2326	.0606	.2606	16.499	3.837	2.576	9.883	**8**
9	5.160	.1938	.0481	.2481	20.799	4.031	2.836	11.434	**9**
10	6.192	.1615	.0385	.2385	25.959	4.192	3.074	12.887	**10**
11	7.430	.1346	.0311	.2311	32.150	4.327	3.289	14.233	**11**
12	8.916	.1122	.0253	.2253	39.581	4.439	3.484	15.467	**12**
13	10.699	.0935	.0206	.2206	48.497	4.533	3.660	16.588	**13**
14	12.839	.0779	.0169	.2169	59.196	4.611	3.817	17.601	**14**
15	15.407	.0649	.0139	.2139	72.035	4.675	3.959	18.509	**15**
16	18.488	.0541	.0114	.2114	87.442	4.730	4.085	19.321	**16**
17	22.186	.0451	.00944	.2094	105.931	4.775	4.198	20.042	**17**
18	26.623	.0376	.00781	.2078	128.117	4.812	4.298	20.680	**18**
19	31.948	.0313	.00646	.2065	154.740	4.843	4.386	21.244	**19**
20	38.338	.0261	.00536	.2054	186.688	4.870	4.464	21.739	**20**
21	46.005	.0217	.00444	.2044	225.026	4.891	4.533	22.174	**21**
22	55.206	.0181	.00369	.2037	271.031	4.909	4.594	22.555	**22**
23	66.247	.0151	.00307	.2031	326.237	4.925	4.647	22.887	**23**
24	79.497	.0126	.00255	.2025	392.484	4.937	4.694	23.176	**24**
25	95.396	.0105	.00212	.2021	471.981	4.948	4.735	23.428	**25**
26	114.475	.00874	.00176	.2018	567.377	4.956	4.771	23.646	**26**
27	137.371	.00728	.00147	.2015	681.853	4.964	4.802	23.835	**27**
28	164.845	.00607	.00122	.2012	819.223	4.970	4.829	23.999	**28**
29	197.814	.00506	.00102	.2010	984.068	4.975	4.853	24.141	**29**
30	237.376	.00421	.00085	.2008	1 181.9	4.979	4.873	24.263	**30**
31	284.852	.00351	.00070	.2007	1 419.3	4.982	4.891	24.368	**31**
32	341.822	.00293	.00059	.2006	1 704.1	4.985	4.906	24.459	**32**
33	410.186	.00244	.00049	.2005	2 045.9	4.988	4.919	24.537	**33**
34	492.224	.00203	.00041	.2004	2 456.1	4.990	4.931	24.604	**34**
35	590.668	.00169	.00034	.2003	2 948.3	4.992	4.941	24.661	**35**
40	1 469.8	.00068	.00014	.2001	7 343.9	4.997	4.973	24.847	**40**
45	3 657.3	.00027	.00005	.2001	18 281.3	4.999	4.988	24.932	**45**
50	9 100.4	.00011	.00002	.2000	45 497.2	4.999	4.995	24.970	**50**
55	22 644.8	.00004	.00001	.2000	113 219.0	5.000	4.998	24.987	**55**
60	56 347.5	.00002		.2000	281 732.6	5.000	4.999	24.994	**60**

Appendix: Solaronics Degree Day Table

TABLE 9

SOLARONICS INC.

OUTSIDE DESIGN TEMPERATURES AND DEGREE DAYS

City	Outside Design Temperature °F		Degree Days Base Temperature - °F									Average Winter Temp.
	99%	97.5%	80	75	70	65	60	55	50	45	40	°F
ALABAMA												
BIRMINGHAM	17	21	6671	5147	3856	2844	1995	1333	838	488	246	54.2
HUNTSVILLE	11	16	7284	5720	4374	3302	2414	1670	1103	686	396	51.3
MOBILE	25	29	4946	3585	2504	1684	1062	619	330	148	61	59.9
MONTGOMERY	22	25	5856	4398	3194	2269	1508	945	547	282	124	55.4
ALASKA												
ANCHORAGE	-23	-18	17586	15095	12736	10911	9122	7492	6081	4896	3860	23.0
FAIRBANKS	-51	-47	20270	18125	16126	14345	12661	11115	9714	8451	7312	6.7
ARIZONA												
FLAGSTAFF	-2	4	13716	11331	9027	7322	5776	4421	3267	2299	1500	35.6
PHOENIX	31	34	5351	3727	2398	1552	899	431	165	45	6	58.5
TUCSON	28	32	4750	3582	2634	1752	1050	541	229	65	17	58.1
WINSLOW			9066	7447	5998	4733	3623	2683	1882	1249	772	
ARKANSAS												
ELDORADO	18	23				2645						
FORT SMITH	12	17	7243	5745	4382	3336	2442	1687	1075	613	306	50.3
LITTLE ROCK	15	20	7417	5843	4417	3354	2442	1687	1075	624	319	50.5
CALIFORNIA												
BAKERSFIELD	30	32	6351	4629	3198	2185	1367	760	371	147	44	55.4
FRESNO	28	30	7124	5322	3764	2650	1724	995	493	205	58	53.3
LOS ANGELES (A)	37	40			3208	1819	833	295	66	7		57.4
RED BLUFF			7060	5309	3770	2688	1762	1018	505	208	60	53.8
SACRAMENTO	30	32	7671	5738	4056	2843	1837	1043	493	186	50	53.9
SAN FRANCISCO	35	38	10890	7490	4762	3042	1668	769	289	67	8	53.4
COLORADO												
DENVER (A)	-5	1	11007	9152	7465	6016	4723	3601	2653	1852	1211	37.6
DENVER (C)			10523	8634	6907	5505	4246	3175	2271	1533	958	40.8
PUEBLO	-7	0	10315	8470	6729	5394	4221	3220	2351	1628	1059	40.4
CONNECTICUT												
BRIDGEPORT	6	9	10244	8465	6845	5461	4264	3216	2321	1583	1003	39.9
HARTFORD	3	7	11435	9959	7811	6350	5085	3971	3005	2173	1507	37.3
DELAWARE												
WILMINGTON	10	14	9265	7669	6226	4940	3824	2839	2003	1330	816	42.5
D.C., WASHINGTON												
(DULLES)			9660	7911	6309	5005	3881	2898	2054	1380	860	
(NATIONAL)	14	17	8250	6742	5383	4211	3182	2293	1563	984	569	45.7
FLORIDA												
JACKSONVILLE	29	32	4373	3085	2090	1327	788	429	194	70	21	61.9
PENSACOLA	25	29	4751	3420	2377	1578	991	575	302	135	56	60.4
TALAHASSEE	27	30	4773	3429	2374	1563	966	550	279	116	48	60.1
TAMPA	36	40	3572	2216	1257	718	364	151	50	14		66.4
GEORGIA												
ATHENS	25	29	6986	5407	4047	2975	2084	1370	830	462	213	51.8
ATLANTA	17	22	7328	5645	4195	3095	2189	1461	911	524	262	51.7
MACON	21	25	5773	4345	3159	2240	1492	934	536	271	111	56.2
SAVANNAH	24	27	5318	3947	2830	1952	1258	751	413	185	72	57.8
IDAHO												
BOISE	3	10	11045	9092	7273	5833	4533	3399	2434	1626	1029	39.7
POCATELLO	-8	-1	12380	10414	8617	7063	5687	4454	3401	2504	1759	34.8
ILLINOIS												
CHICAGO (O'HARE)	-8	-4	11425	9601	7942	6497	5245	4163	3220	2404	1742	35.8
CHICAGO (MIDWAY)	-5	0	10875	9133	7486	6127	4952	3912	2998	2219	1574	37.5
MOLINE	-9	-4	10946	9275	7758	6395	5202	4170	3263	2462	1816	36.4
PEORIA	-8	-4	10621	8955	7432	6098	4930	3910	3013	2239	1612	38.1
ROCKFORD	-9	-4	11588	9850	8274	6845	5600	4507	3555	2713	2011	34.8
SPRINGFIELD	-3	2	10022	8364	6848	5558	4437	3468	2615	1913	1324	40.6
INDIANA												
EVANSVILLE	4	9	9029	7355	5838	4624	3578	2685	1929	1327	857	45.0
FORT WAYNE	-4	1	11111	9296	7605	6209	4992	3930	2996	2193	1537	37.3
INDIANAPOLIS	-2	2	10256	8503	6887	5577	4430	3431	2568	1856	1277	39.6
SOUTH BEND	-3	1	11450	9605	7899	6462	5213	4118	3156	2333	1656	36.6
TERRE HAUTE	-2	4				5351						

City	Outside Design Temperature °F 99%	Outside Design Temperature °F 97.5%	Degree Days Base Temperature - °F 80	75	70	65	60	55	50	45	40	Average Winter Temp. °F
IOWA												
BURLINGTON	-7	-3	10636	8984	7479	6149	4988	3970	3077	2308	1681	37.6
DES MOINES	-10	-5	11170	9545	8082	6710	5521	4470	3546	2728	2045	35.5
DUBUQUE	-12	-7	10917	9277	8761	7277	5992	4871	3893	3028	2286	32.7
SIOUX CITY	-11	-7	11431	9800	8329	6953	5745	4674	3738	2898	2196	34.0
WATERLOO	-15	-10	12015	10347	8821	7433	6180	5059	4065	3194	2442	32.6
KANSAS												
CONCORDIA			10136	8466	6911	5623	4498	3509	2646	1912	1320	40.4
DODGE CITY	0	5	9280	7716	6297	5046	3963	3011	2184	1512	962	42.5
GOODLAND	-5	0	11105	9262	7508	6119	4891	3804	2857	2041	1387	37.8
RUSSELL	0	4	9538	7976	6561	5312	4220	3259	2423	1735	1167	
SALINA	0	5				4992						
TOPEKA	0	4	9656	8017	6492	5243	4152	3203	2378	1700	1131	41.7
WICHITA	3	7	8831	7287	5876	4687	3654	2750	1977	1346	842	44.2
KENTUCKY												
BOWLING GREEN	4	10										
LEXINGTON	3	8	9293	7566	5974	4729	3652	2743	1963	1350	879	43.8
LOUISVILLE	5	10	9241	7482	5866	4640	3584	2676	1906	1303	836	44.0
LOUISIANA												
ALEXANDRIA	23	27	5780	4328	3118	2200	1443	880	490	234	94	57.5
BATON ROUGE	25	29	4983	3603	2495	1670	1036	582	295	120	45	59.8
LAKE CHARLES	27	31	4681	3345	2291	1498	908	500	240	91	34	60.5
MONROE	20	25										
NEW ORLEANS	29	33	4675	3315	2251	1465	893	492	239	96	36	61.0
SHREVEPORT	20	25	5596	4220	3067	2167	1438	883	490	233	88	56.2
MAINE												
CARIBOU	-18	-13	15370	13269	11360	9632	8044	6634	5409	4319	3379	24.4
OLD TOWNE			14414	12293	10321	8648	7133	5800	4628	3589	2704	
PORTLAND	-6	-1	13251	11121	9141	7498	6035	4764	3658	2705	1908	33.0
MARYLAND												
BALTIMORE	10	13	9314	7577	5976	4729	3631	2682	1873	1236	753	43.7
CUMBERLAND	6	10										
MASSACHUSETTS												
BLUE HILL			11692	9698	7852	6335	5020	3885	2895	2071	1406	
BOSTON	6	9	10801	8849	7049	5621	4383	3313	2405	1659	1081	40.0
NANTUCKET			11464	9395	7546	5929	4520	3323	2311	1513	912	40.2
WORCESTER	0	4	12218	10236	8403	6848	5498	4326	3296	2421	1697	34.7
MICHIGAN												
ALPENA	-11	-6	14067	12034	10191	8518	6982	5635	4473	3464	2586	29.7
DETROIT (A)			11384	9564	7864	6419	5167	4072	3113	2280	1597	37.1
DETROIT (C)	3	6	11089	9302	7639	6228	5004	3936	2991	2178	1506	37.2
FLINT	-4	1	12070	10235	8581	7041	5705	4540	3529	2640	1892	33.1
GRAND RAPIDS	1	5	11713	9925	8288	6801	5514	4383	3396	2524	2213	34.9
LANSING	-3	1	11836	10037	8404	6904	5608	4464	3470	2595	1860	34.8
MARQUETTE	-12	-8	13727	11767	10015	8351	6835	5517	4379	3378	2507	30.2
MUSKEGON	2	6	12008	10138	8421	6890	5550	4390	3373	2482	1732	36.0
SAULT STE. MARIE	-12	-8	14679	12681	10913	9193	7614	6215	5017	3971	3050	27.7
TRAVERSE CITY	-3	1	12931	11022	9294	7698	6272	5035	3953	3013	2200	
MINNESOTA												
BEMIDJI	-31	-26				10203						
DULUTH	-21	-16	15320	13278	11450	9756	8185	6793	5581	4540	3625	23.4
INTERNAT'L FALLS	-29	-25	16048	14037	12237	10547	8995	7623	6413	5348	4412	
MINNEAPOLIS	-16	-12	12815	11128	9657	8159	6842	5677	4668	3765	2960	28.3
ROCHESTER	-17	-12	13029	11285	9768	8227	6868	5682	4643	3739	2916	28.8
ST. CLOUD	-15	-11	13745	11976	10427	8868	7481	6255	5187	4241	3385	
MISSISSIPPI												
JACKSON	21	25	5855	4407	3206	2300	1548	988	590	319	147	55.7
MERIDIAN	19	23	5978	4526	3315	2388	1621	1042	623	339	159	55.4
MISSOURI												
COLUMBIA REG	-1	4	4312	7739	6329	5078	3997	3064	2259	1605	1058	42.3
KANSAS CITY	2	6	9520	7892	6390	5161	4089	3157	2351	1694	1140	43.9
ST. JOSEPH	-3	2	9901	8235	6696	5435	4341	3378	2544	1847	1275	40.3
ST. LOUIS	2	6	9116	7464	5959	4750	3701	2798	2031	1419	925	43.1
SPRINGFIELD	3	9	8844	7241	5799	4570	3517	2611	1844	1235	745	44.5
MONTANA												
BILLINGS	-15	-10	12911	10781	8799	7265	5898	4697	3641	2766	2034	34.5
CUT BANK	-25	-20	15271	12918	10748	9033	7474	6096	4907	3886	3084	
DILLON	-18	-10	14395	12155	10031	8354	6821	5457	4255	3237	2375	
GLASGOW	-22	-18	14292	12315	10533	8969	7572	6329	5238	4302	3481	26.4
GREAT FALLS	-21	-15	13464	11256	9250	7652	6248	5022	3965	3074	2334	32.8
HAVRE	-22	-15	14091	12068	10267	8687	7282	6073	5005	4104	3304	28.1

City	Outside Design Temperature °F		Degree Days Base Temperature - °F									Average Winter Temp.
	99%	97.5%	80	75	70	65	60	55	50	45	40	°F
MONTANA, Cont'd.												
HELENA	-21	-16	14118	11907	9834	8190	6710	5389	4247	3258	2419	31.1
KALISPELL	-14	-7	14972	12580	10281	8554	6959	5542	4304	3233	2350	31.4
LEWISTON	-22	-16	14794	12462	10270	8586	7038	5676	4487	3467	2620	
MILES CITY	-20	-15	13010	11105	9353	7889	6552	5392	4369	3479	2706	31.2
MISSOULA	-13	-6	13085	10852	9609	7931	6410	5066	3884	2876	2039	31.5
NEBRASKA												
GRAND ISLAND	-8	-3	11014	9330	7772	6420	5224	4166	3239	2434	1761	36.0
LINCOLN	-5	-2	10677	9041	7539	6218	5062	4040	3139	2362	1717	38.8
NORFOLK	-8	-4	11605	9913	8362	6981	5745	4663	3710	2863	2156	34.0
NORTH OMAHA			11131	9468	7963	6601	5400	4349	3427	2624	1954	
NORTH PLATTE	-8	-4	11717	9884	8161	6743	5470	4345	3354	2509	1787	35.5
OMAHA	-8	-3	10335	8746	7346	6049	4907	3911	3037	2290	1669	35.6
SCOTTSBLUFF	-8	-3	11983	10065	8235	6774	5473	4304	3289	2415	1684	35.9
VALENTINE			12179	10398	8752	7300	6006	4859	3847	2956	2193	32.6
NEVADA												
ELKO	-8	-2	13114	11033	9103	7483	6027	4714	3586	2625	1827	34.0
ELY	-10	-4	13340	11325	9473	7814	6327	5004	3826	2829	1977	33.1
LAS VEGAS	25	28	6196	4785	3541	2601	1770	1120	625	306	112	53.5
RENO	5	10				6022						39.3
NEW HAMPSHIRE												
BERLIN	-14	-9										
CONCORD	-8	-3	12795	10834	8947	7360	5967	4757	3682	2762	1978	33.0
MANCHESTER	-8	-3				7101						
MT. WASHINGTON			20571	18187	15703	13878	12053	10253	8534	6960	5585	15.2
NEW JERSEY												
NEWARK	10	14	9929	8082	6327	5034	3911	2920	2074	1391	863	42.8
TRENTON	11	14	9614	7859	6248	4947	3818	2832	1996	1323	810	42.4
NEW MEXICO												
ALBUQUERQUE	12	16	8424	6892	5504	4292	3234	2330	1557	963	522	45.0
CLAYTON			10006	8218	6610	5207	3999	2966	2089	1374	821	42.0
ROSWELL	13	18	7979	6339	4836	3697	2729	1898	1226	706	364	47.5
TRUTH OR CONSEQU.	14	18	7339	5845	4520	3392	2447	1636	1007	542	258	
NEW YORK												
ALBANY	-6	-1	11964	10093	8375	6888	5596	4451	3457	2595	1885	34.6
BINGHAMTON	-2	1	12557	10618	8855	7285	5908	4714	3677	2767	2009	36.6
BUFFALO	2	6	12170	10249	8464	6927	5591	4429	3403	2508	1766	34.5
ELMIRA	-4	1										
MASSENA	-13	-8	13501	11570	9828	8237	6827	5596	4510	3552	2733	
NEW YORK - JFK	12	15	9903	8140	6525	5184	4023	2994	2130	1422	880	41.4
NEW YORK - LaGuard.	11	15	9484	7767	6192	4909	3787	2806	1980	1311	798	43.1
OSWEGO	1	7	11852	10014	8340	6792	5544	4274	3243	2376	1636	
ROCHESTER	1	5	11827	9948	8217	6719	5417	4285	3291	2434	1715	35.4
SYRACUSE	-3	2	11738	9868	8164	6678	5379	4250	3267	2429	1725	35.2
NORTH CAROLINA												
ASHEVILLE	10	14	9045	7179	5548	4237	3129	2224	1488	937	549	46.7
CHARLOTTE	18	22	7425	5760	4328	3218	2300	1552	984	585	308	50.4
GREENSBORO	14	18	8171	6479	4997	3825	2811	1984	1324	825	483	47.5
RALEIGH/DURHAM	16	20	7863	6155	4662	3514	2542	1744	1123	670	367	49.4
WILMINGTON	23	26	6243	4694	3411	2433	1632	1028	610	321	142	54.6
NORTH DAKOTA												
BISMARCK	-23	-19	14161	12283	10596	9044	7656	6425	5326	4374	3542	26.6
FARGO	-22	-18	14237	12431	10816	9271	7891	6663	5573	4615	3753	24.8
MINOT	-24	-20	14673	12740	11005	9407	7964	6685	5564	4573	3717	24.9
WILLISTON	-25	-21	14402	12470	10735	9161	7753	6504	5387	4450	3598	25.5
OHIO												
AKRON/CANTON	1	6	11448	9492	7675	6224	4971	3883	2936	2126	1482	38.1
CINCINNATI (A)	1	6	9754	7978	6353	5070	3965	3001	2189	1527	1004	41.4
CINCINNATI (C)	1	6	9434	7683	6088	4844	3763	2830	2040	1412	921	45.1
CLEVELAND	1	5	11329	9395	7610	6154	4901	3819	2876	2079	1433	37.2
COLUMBUS (A)	0	5	10730	8835	7068	5702	4513	3480	2597	1846	1260	39.7
DAYTON	-1	4	10463	8653	6970	5641	4483	3468	2600	1866	1282	39.8
MANSFIELD	0	5	10824	8943	7197	5818	4618	3573	2679	1917	1317	36.9
TOLEDO	-3	1	11395	9541	7815	6381	5136	4049	3091	2274	1602	36.4
YOUNGSTOWN	-1	4	11698	9735	7922	6426	5145	4032	3054	2232	1558	36.8
OKLAHOMA												
OKLAHOMA CITY	9	13	7662	6151	4796	3695	2760	1962	1326	804	455	48.3
TULSA	8	13	7594	6113	4759	3680	2750	1950	1306	778	418	47.7

City	Outside Design Temperature °F 99%	97.5%	Degree Days Base Temperature - °F 80	75	70	65	60	55	50	45	40	Average Winter Temp. °F
OREGON												
EUGENE	17	22	10732	8432	6378	4739	3313	2141	1226	607	255	45.6
MEACHAM			13978	11729	9598	7863	6249	4817	3556	2495	1620	34.2
MEDFORD	19	23	10095	8167	6426	4930	3614	2496	1577	882	419	43.2
NORTH BEND	29	32	12846	9509	6504	4688	2985	1642	756	292	84	
PENDLETON	-2	5	10297	8399	6698	5240	3968	2868	1970	1264	779	42.6
PORTLAND	17	23	10696	8424	6407	4792	3385	2234	1333	708	338	45.6
PENNSYLVANIA												
ALLENTOWN	4	9				5827						38.9
BRADFORD			15118	11367	9478	7804	6294	5006	3894	2931	2139	
ERIE	4	9	12317	10286	8419	6851	5485	4304	3283	2411	1684	36.8
HARRISBURG	7	11	9759	8067	6517	5224	4087	3097	2238	1541	990	41.2
PHILADELPHIA	10	14	9470	7731	6125	4865	3753	2788	1965	1312	809	41.8
PITTSBURGH (A)	1	5	11127	9172	7366	5930	4694	3637	2720	1938	1326	38.4
PITTSBURGH (C)	3	7	10112	8276	6591	5278	4135	3138	2294	1603	1065	42.2
SCRANTON/ WILKES-BARRE	1	5	11188	9379	7735	6277	5018	3928	2972	2149	1486	37.2
WILLIAMSPORT	2	7	11044	9162	7404	5981	4757	3695	2764	1971	1341	38.5
RHODE ISLAND												
BLOCK ISLAND			11287	9216	7340	5771	4432	3289	2306	1517	917	40.1
PROVIDENCE	5	9	11083	9181	7458	5972	4682	3565	2599	1803	1172	38.8
SOUTH CAROLINA												
CHARLESTON	25	28	5687	4246	3070	2146	1406	864	496	240	102	56.4
COLUMBIA	20	24	6381	4867	3592	2598	1783	1154	686	374	165	54.0
GREENVILLE/ SPARTANBURG	18	22	7234	5646	4276	3163	2246	1493	921	519	250	
SOUTH DAKOTA												
ABERDEEN	-19	-15	13500	11718	10129	8617	7267	6078	5014	4087	3272	
HURON	-18	-14	12780	11057	9513	8055	6751	5600	4582	3678	2889	28.8
PIERRE	-15	-10	12424	10681	9095	7677	6401	5271	4273	3409	2647	
RAPID CITY	-11	-7	12589	10634	8804	7324	5982	4799	3762	2868	2115	33.4
SIOUX FALLS	-15	-11	12507	10811	9296	7838	6543	5401	4394	3498	2712	30.6
TENNESSEE												
BRISTOL	13	18	8905	7122	5550	4306	3255	2373	1646	1093	687	46.2
CHATTANOOGA	13	18	7735	6068	4610	3505	2574	1785	1180	737	433	50.3
KNOXVILLE	13	19	7627	5997	4590	3478	2557	1775	1187	744	439	49.2
MEMPHIS	13	18	7123	5601	4260	3227	2352	1624	1058	640	349	50.5
NASHVILLE	9	14	7802	6210	4817	3696	2758	1964	1338	852	518	48.9
OAK RIDGE			8343	6632	5132	3944	2955	2119	1445	933	576	47.7
TEXAS												
ABILENE	15	20	6165	4774	3560	2610	1801	1162	664	342	139	53.9
AMARILLO	6	11	8544	6891	5399	4183	3156	2278	1548	976	555	47.0
AUSTIN	24	28	4968	3643	2560	1737	1097	620	316	127	44	59.1
DALLAS	18	20	5715	4358	3190	2290	1544	949	526	250	94	53.3
DEL RIO	26	31	4632	3339	2284	1523	923	494	230	80	23	
EL PASO	20	24	6469	4974	3670	2678	1833	1149	653	326	129	52.9
FORT WORTH	17	22	5879	4496	3304	2382	1616	1007	562	274	105	55.1
GALVESTON (C)	31	36	4285	2964	1947	1224	704	369	157	54	18	62.0
HOUSTON	27	32	4620	3274	2210	1434	864	471	215	81	26	61.0
LUBBOCK	10	15	7626	6069	4661	3545	2603	1807	1163	666	336	48.8
LUFKIN	25	29	5251	3912	2796	1940	1253	731	385	163	53	
SAN ANGELO	18	22	5653	4301	3138	2240	1498	918	493	227	83	56.0
SAN ANTONIO	25	30	4905	3503	2374	1570	956	518	242	92	25	60.1
WACO	20	25	5526	4132	2985	2058	1357	807	437	195	62	57.2
WICHITA FALLS	14	18	6534	5141	3936	2904	2061	1384	832	451	192	53.0
UTAH												
SALT LAKE CITY	3	8	10720	8976	7362	5983	4733	3633	2676	1864	1220	38.4
WENDOVER			10483	8714	7096	5760	4558	3511	2621	1870	1281	39.1
VERMONT												
BURLINGTON	-12	-7	13140	11207	9443	7876	6488	5270	4190	3246	2442	29.4
VIRGINIA												
LYNCHBURG	12	16	8533	6909	5471	4233	3172	2269	1536	966	570	46.0
NORFOLK	20	22	7847	6127	4635	3488	2516	1710	1100	663	366	49.2
RICHMOND	14	17	8555	6751	5119	3939	2916	2061	1388	866	508	47.3
ROANOKE	12	16	8762	7073	5595	4315	3227	2321	1586	1010	580	46.1
WALLOPS ISLAND			8770	7043	5540	4250	3161	2262	1538	976	561	
WASHINGTON												
BELLINGHAM	10	15				5738						
OLYMPIA	16	22	11747	9397	7267	5530	3970	2653	1617	854	396	44.2
QUILLAYUTE			12828	10209	7765	5951	4232	2750	1603	813	354	

City	Outside Design Temperature °F		Degree Days Base Temperature - °F									Average Winter Temp.
	99%	97.5%	80	75	70	65	60	55	50	45	40	°F
WASHINGTON, Cont'd.												
SEATTLE	22	27	10887	8366	6148	4487	3046	1893	1045	501	197	46.9
SEATTLE/TACOMA	21	26	11516	9085	6902	5185	3657	2386	1416	731	325	44.2
SPOKANE	-6	2	12438	10353	8423	6835	5420	4173	3088	2188	1464	36.5
STAMPEDE PASS			15847	13510	11209	9400	7643	6006	4532	3256	2203	
WALLA WALLA	0	7	9748	7882	6224	4835	3616	2600	1760	1126	685	43.8
YAKIMA	-2	5	11673	9522	7551	6009	4655	3483	2502	1688	1084	39.1
WEST VIRGINIA												
BECKLEY			10883	8879	7106	5615	4356	3279	2390	1652	1093	
CHARLESTON	7	11	9301	7487	5858	4590	3500	2590	1809	1216	766	44.9
ELKINS	1	6	11603	9453	7525	5975	4659	3533	2616	1834	1245	40.1
HUNTINGTON	5	10	9427	7570	5889	4624	3533	2624	1843	1249	797	45.0
PARKERSBURG	7	11	9486	7701	6098	4817	3720	2786	1987	1363	885	43.5
WISCONSIN												
EAU CLAIRE	-15	-11	13176	11443	9934	8388	7033	5832	4786	3860	3028	
GREEN BAY	-13	-9	13043	11246	9678	8098	6689	5473	4405	3478	2644	30.3
LA CROSSE	-13	-9	11975	10323	8868	7417	6158	5050	4088	3219	2458	31.5
MADISON	-11	-7	12644	10855	9274	7730	6373	5188	4156	3250	2447	30.9
MILWAUKEE	-8	-4	12535	10671	8989	7444	6080	4898	3860	2946	2176	32.6
WYOMING												
CASPER	-11	-5	12985	10998	9106	7555	6167	4914	3813	2857	2041	33.4
CHEYENNE	-9	-1	12942	10849	8859	7255	5825	4562	3452	2512	1717	34.2
LANDER	-16	-11	13390	11354	9439	7869	6471	5207	4080	3140	2321	31.4
ROCK SPRINGS	-9	-3	14142	12033	10059	8410	6922	5592	4412	3393	2533	
SHERIDAN	-14	-8	13353	11251	9286	7708	6298	5024	3935	3000	2200	32.5

Index

About the Author

George Kritzler graduated with a B.S. in Mechanical Engineering from Newark College of Engineering and then an M.B.A. from Fairleigh Dickinson University. His work experience was as a Sales Engineer with the Public Service Electric and Gas Company. This was his first design experience using gas IR heating to compete with #2 oil. Over 50 installations of gas IR heating systems were designed and sold while with PSE&G.

During his short career as Industrial and Commercial Sales Manager with Elizabethtown Gas Company, his team of 12 Sales Engineers sold at least 12 gas IR systems. A severe natural gas shortage during the early 1970s shortened his career there with no gas to sell.

As Senior HVAC Engineer with Consolidated Edison Co. of NY in the Plant Structures Department, Mr. Kritzler was surprised to find that none of the four HVAC engineers knew about gas IR heating. That ended when he showed them the advantages of gas IR heating. He personally designed and supervised the installation of over 40 gas IR heating systems.

In retirement, he was able to become an Energy Conservation Engineer/Consultant by establishing Imagineers Unltd. Two of the buildings that successfully followed his recommendations were the New York Academy of Medicine and the Riverside Church. These were approved by the New York State Energy Research & Development Authority's energy reduction program with annual savings of $26,800 and $48,000, respectively.

For Product Safety Concerns and Information please contact our
EU representative GPSR@taylorandfrancis.com Taylor & Francis
Verlag GmbH, Kaufingerstraße 24, 80331 München, Germany